Enzyme Chemistry

Impact and applications

Contents

Preface

In the molecular sciences, enzyme chemistry occupies a special niche as one of the major contact points between chemical and biological disciplines. The special properties of enzymes as selective and efficient catalysts are so central to current challenges to chemists that the development of enzyme chemistry in the past thirty years has been a major stimulus to chemical research in general. On the one hand studies of the intrinsic properties of enzymes and, on the other hand, their applications to synthesis, drug design, and biosynthesis have had an immense impact. This book brings together in one volume essays describing several such fields with emphasis on the applications. It would be unnecessarily repetitious to outline the approach and contents of the book in a Preface; the first short chapter is more eloquent than a formal Preface can be. I shall therefore encourage you to begin with the Introduction in Chapter 1 and here I wish to extend my warm thanks to those who have contributed to the production of this book: the authors for their acceptance of the overall concept of the book and for the thoughtfulness of their writing; Dr Charles Suckling, FRS and Professor Hamish Wood for their constructive criticism of the whole book; and Dr John Buckingham and his colleagues at Chapman and Hall for their efficiency and enthusiasm in transforming the typescripts into the book that you now hold.

Colin J. Suckling
University of Strathclyde

Contributors

Donald H. Brown   Department of Pure and Applied Chemistry, University of Strathclyde, Glasgow, UK

David E. Cane   Department of Chemistry, Brown University, Rhode Island, USA

Barrie Hesp   Medicinal Chemistry, Stuart Pharmaceuticals, Division of ICI Americas Inc., Wilmington, USA

Ronald Kluger   Department of Chemistry, University of Toronto, Ontario, Canada

Seiji Shinkai   Department of Industrial Chemistry, Nagasaki University, Nagasaki, Japan

W. Ewen Smith   Department of Pure and Applied Chemistry, University of Strathclyde, Glasgow, UK

Colin J. Suckling   Department of Pure and Applied Chemistry, University of Strathclyde, Glasgow, UK

Keith E. Suckling   Department of Biochemistry, University of Edinburgh Medical School, Edinburgh, UK

Alvin K. Willard   Medicinal Chemistry, Stuart Pharmaceuticals, Division of ICI Americas Inc., Wilmington, USA

1 | Infant enzyme chemistry

Colin J. Suckling

When this book was first planned, the idea in mind was to review, through a series of personal but related essays, the major impact that the study of enzymes has had upon some important fields of chemistry in the last thirty years. It was therefore something of a surprise to discover in the nineteenth century literature that enzymes had already prompted a great deal of chemical research, some of it with a remarkably modern ring, as I shall try to show in the next few pages. As early as 1833 observations had been made of the phenomenon of the natural hydrolysis of potato starch but with vitalistic concepts still much in people's minds, it was difficult to accept the existence of biological catalysts. The idea that enzymes are chemicals provoked prolonged scepticism and controversy. During the first half of the nineteenth century further naturally occurring reactions were recognized, in particular fermentations involving yeasts. On the one hand, it was held that the enzymic activity responsible for these fermentations was a property inseparable from living cells. Pasteur, amongst others, took this view. On the other hand, Liebig and, not surprisingly, Wöhler, regarded enzymes as chemical catalysts, albeit of unknown constitution, that could be separated from cells. Indeed these two may well have conspired to lampoon vitalism in an anonymous paper in Liebig's *Annalen der Pharmacie* (Anon., 1839). In this amusing article we read of chemical reactions brought about by

'small animals which hatch from eggs (yeast) in sugary solution and which on microscopic examination are seen to take the form of a Beindorf distillation apparatus, without the condenser . . . these animals, which have neither teeth nor eyes, but possess a stomach, a bladder which, when full, looks like a champagne bottle . . . devour the sugar with the production of excrement as alcohol and carbon dioxide.'

Eventually the argument was settled by experiment. In 1897, Buchner demonstrated that a yeast extract was capable of sustaining the fermentation of sugar but a few years earlier, a remarkable series of contributions began to appear from the laboratory of Emil Fischer (1894). The papers make enthralling reading, not only for their scientific content, but also because they convey great enthusiasm, sometimes naive, but always evident. The main subject to which Fischer addressed his powerful experimental skills and penetrating intellect was stereoselectivity in enzymic catalysis, a field still of current significance; the ability of enzymes to select between stereoisomers has proved one of their most alluring properties.

Fischer's paper (1894) is remarkable for its discoveries themselves and also for the insight of a man of genius into future developments. He was, of course, uniquely well placed to tackle the problem of stereoselectivity because he had available an extensive series of stereoisomeric sugars which he had synthesized to determine their configurations. Derivatives of these compounds served as substrates for glycosidases which even in those days were available in crude cell-free form. His paper begins

'The different properties of the stereoisomeric hexoses with respect to yeast led Thierfelder and I to the hypothesis that the active chemical agent of yeast cells can only attack those sugars to which it possesses a related configuration.'

The hypothesis was supported by demonstrating, amongst other things, that the enzyme that hydrolyses sucrose, called 'invertin' by Fischer, acts only upon α-D-glucosides: β-D-glucosides and L-glucosides were completely untouched. There was no doubt that this was not just a chance phenomenon because a second enzyme, emulsin, was found to hydrolyse β-D-glucosides of both synthetic and natural origin. The complementary nature of these results is conclusive and, of course, still important in modern stereochemical studies. Fischer's assessment of his results is fascinating reading. It also makes an admirable preface to this book because it foreshadows much of what follows. When you have read further, you may be interested to reflect upon these lines:

'But the results suffice in principle to show that enzymes are choosy with respect to the configuration of their substrate, like yeast and other micro-organisms. The analogy between both phenomena appears so complete in this respect that one may assume the same origin for them, and accordingly, I return to the abovementioned hypothesis of Thierfelder and myself. Invertin and emulsin have many perceptible similarities and consist doubtless of an asymmetrically built molecule . . . To use an image, I would say that the enzyme and glucoside must fit each other like a lock and key to be able to exert a chemical influence upon each other . . . The facts proven for the complex enzymes will soon also be found with simpler asymmetric agents. I scarcely doubt that enzymes will be of use for the determination of configuration of asymmetric substances . . . The earlier much accepted distinction between the chemical ability of living cells and the action of chemical agents with regard to molecular asymmetry does not in fact exist.'

Although the last sentence quoted was directed to his contemporaries, much of the preceding extract reads remarkably freshly to modern chemists nearly a century later. We have the advantage over Fischer in techniques, but some of the concepts that he advanced have still to be realized in perfection as we shall see. However, Fischer was by no means the only scientific prophet in the field of stereochemistry and his work depended much upon the understanding developed by Pasteur. There is little in modern stereochemical research that does not derive something from the experimental and conceptual contribution of these two great scientists (see Robinson, 1974).

It is remarkable how much was achieved in Fischer's time with impure enzyme preparations. A parallel in today's research might be the study of preparations containing unpurified neurotransmitter or hormone receptors, although these too are now amenable to purification by modern chromatographic techniques. As Fischer predicted, enzymes have become widely used for the determination of configuration but it is only in recent years that 'simpler asymmetric agents' have been able to reproduce enzymic stereoselectivity (see Chapters 3 and 4). Not surprisingly, the ever enthusiastic Fischer even had a go at asymmetric synthesis himself (Fischer and Slimmer, 1903, and see Fig. 1.1). Knowing that glucose is chiral, Fischer hoped that the naturally occurring glycoside, helicin, would undergo asymmetric addition at the carbonyl group guided in some way by the asymmetric environment created by the glucose ring. This strategy has since proved successful (Chapter 4) and had Fischer used a more bulky nucleophile, he too might have been successful. His first attempt was to add hydrogen cyanide to helicin and to hydrolyse the product carefully. An optically inactive product resulted. So Fischer tried again using diethyl zinc and this time the product obtained from vacuum distillation was optically active. In the exhilaration of discovery he wrote '. . . with this we thus believed that we had solved the problem of asymmetric synthesis.'! Then came the snag, Gilbertian 'modified rapture'. Rigorous control experiments clearly showed that the apparent asymmetric induction was due to an impurity derived from glucose during distillation and no further attempts were reported. Many people have had similar, but unpublished, experiences.

The turn of the century also marked the first steps in the synthetic use of

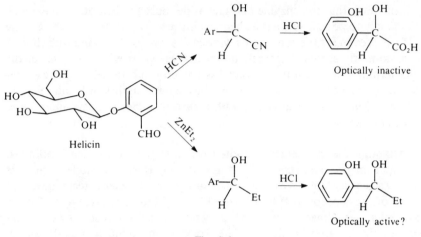

Fig. 1.1

enzymes. Croft-Hill (1898) demonstrated that yeast enzymes could be used synthetically and Emmerling (1900) reported a synthesis of the glycoside amygdalin using enzymes. These, and other pioneering contributions, are cited by Hoesch (1921) in a special edition of *Berichte* devoted entirely to a biography of Fischer. Many of Hoesch's comments are equally apt today more than sixty years later. For instance, in summarizing Fischer's contribution to enzyme chemistry, Hoesch remarks 'Pure chemists may certainly not feel at home with the enzymatic studies of Emil Fischer'. Another notable comment was that Fischer's lock and key metaphor describing enzymic specificity was much appreciated in his day. From Hoesch's review and Fischer's own writings, it seems possible that Fischer never intended this image to be a scientific hypothesis but used it to illuminate the concept of stereochemical biospecificity to an audience totally unfamiliar with the new idea. Modern work, of course, makes it clear that the physical rigidity of a lock and key do not make an appropriate description of a conformationally mobile enzyme–substrate interaction. Once he had demonstrated bio-specificity with enzymes, similar complementary interactions were enthusiastically discussed for the behaviour of other biosystems such as toxins. However naive the metaphor, it was certainly seminal.

Yet another part of our story began in the 1890s. Scientists were not only studying microbial enzymes but mammalian systems were also beginning to be investigated. In 1898, the kidney was shown to contain proteolytic activity (Tigerstedt and Bergmann, 1898). It was further demonstrated that an enzyme named renin hydrolyses a large plasma peptide, which today we know as angiotensinogen, to angiotensin I. We now know that angiotensin I has very little activity in the central or peripheral nervous system; it is further hydrolysed to a smaller peptide, angiotensin II, by an enzyme known as angiotensin converting enzyme. Angiotensin II has powerful effects on the circulatory system and studies of inhibitors of this enzyme have recently developed into one of the classics of modern drug invention (see Chapter 5).

Although much current work was foreshadowed or even initiated at the turn of the century, yet from that time, chemists' contact with enzymes became more remote as for the next five decades, chemists, with some notable exceptions, pursued the systematic study of the reactivity of organic compounds. Sir Robert Robinson was one such exception. Whilst contributing greatly to natural product chemistry and of course to ideas concerning reactivity, he realized that enzymes catalyse reactions under very mild conditions and sought laboratory analogues in syntheses of alkaloids (Robinson, 1917). Meanwhile, biochemists wanted to find out in detail what enzymes are and set about their purification. The first systematic attempts were begun by Willstaetter in the 1920s but the first substantive success came from Sumner who in 1926 reported the crystallization of urease. Perhaps because he couldn't believe that someone else had done it first, Willstaetter disputed that Sumner actually had an enzyme. Nevertheless, proteolytic enzymes were soon purified to crystallinity and it became clear that enzymes are, as Fischer had surmised, proteins.

Although purified enzymes were available from that time on, chemists were by no means ready to accept the idea of macromolecules, let alone macromolecular catalysts. Staudinger, one of the fathers of polymer chemistry, had great difficulty in persuading the Swiss Chemical Society, at a meeting that ended in uncharacteristic Swiss uproar, that macromolecules can exist. A similar scepticism greeted the ideas of a young physical chemist, McBain, concerning the nature of micelles at a meeting of The Royal Society in London. He was told that his notions of molecular aggregation were 'nonsense'. In Germany too Hans Fischer, who established the structures of porphyrins by degradation and synthesis, as late as 1937 appeared to be unaware of the wide physiological importance of porphyrins although the isolation of the porphyrin-containing proteins, cytochromes, had been described in the mid 1920s by Keilin. Despite their temporary but acute myopia with regard to enzymes, chemists at this time were making great strides in understanding the basis of mechanistic organic chemistry. In time, the synthesis of artificial polymers was demonstrated and natural macromolecules too became respectable. The conceptual basis for a symbiotic growth of organic chemistry and enzyme chemistry was founded. This book recounts some of the branches of this growth.

What in particular amongst the properties of enzymes has been most significant for chemistry? In the first place, enzymes are such excellent catalysts. Indeed it has been argued that enzymes have evolved to perfect their catalytic function (see Chapter 2). If this is so, then it is a formidable challenge for chemists to understand the chemical basis for enzymic catalysis and a still greater one to mimic it effectively. However, in addition to these purely scientific aims, there are also extremely important practical consequences of the properties of enzymes. Selectivity in catalysis, as Fischer surmised, is one of the most important and it can be applied in a direct sense to perform both regioselective and stereoselective transformations in organic synthesis (Chapter 4). In addition, selective enzyme inhibitors are immensely important as drugs for the treatment of bacterial and viral diseases (Chapter 5).

In the last twenty years, great stides have been made in our understanding of the chemical basis of enzymic catalysis and it is the application of this and related enzyme chemistry that is developing apace. In the chapters that follow, a team of authors from many different countries and backgrounds discuss enzyme chemistry in relation to two broad themes, firstly synthetic organic chemistry. You will read how the study of the mechanism of action of coenzymes has led to a number of novel synthetic reactions (Chapter 3, Seiji Shinkai). Coenzymes are a good starting point for organic chemists because they are relatively small molecules with some innate catalytic activity even in the absence of enzymes. The wide range of reactivity observable is in itself fascinating and Professor Shinkai reviews much recent work for the first time. How conventional synthetic reactions compare in selectivity with enzyme-catalysed reactions and biomimetic systems is the topic of my own contribution (Chapter 4). The second theme, by way of contrast, concerns more biologically significant topics. The relevance of enzyme chemistry to chemotherapy in organic and inorganic aspects is discussed respectively in Chapters 5 (Barrie Hesp and Alvin Willard) and 6 (Donald Brown and Ewen Smith). Biosynthetic studies have always built a bridge between organic chemistry and biochemistry and recent developments in this field are reviewed by David Cane in Chapter 7. Enzyme chemistry has borrowed much from and given much to biochemistry and the book ends with a consideration of future interactions between the disciplines (Chapter 8, Keith Suckling). To provide a basis for these discussions and to demonstrate the depth of thought into catalysis itself that enzyme properties have provoked Ron Kluger begins the essays with some thoughts on the mechanistic basis of enzyme catalysis (Chapter 2). Interestingly and, to some extent, coincidentally several subject areas are discussed from different points of view by several contributors. These topics include enzyme stereochemistry (Chapters 2–5), prostaglandin chemistry (Chapters 4 and 5), β-lactam antibiotics (Chapters 4, 5 and 7), cyclodextrins (Chapters 2–4), and genetic engineering (Chapters 4, 7 and 8). Further areas of chemistry could also have been selected but these seven essays will give the reader insight into the impact of enzyme chemistry upon laboratory and industrial chemistry and the contacts of chemistry with the life sciences.

Now is the time to let each author speak for himself. In editing this book, I have learned much from the thoughts of my fellow contributors. I am sure that they will convince you too that enzyme chemistry has contributed much to chemistry and is still vibrant and vital. Whilst the subject continues to develop, new challenges for biological chemistry are emerging, challenges that can be met all the better because of what the chemist has learned from enzymes in methods, concepts and techniques. As was alluded to earlier, other proteins can now be purified, in particular antibodies and neurotransmitter and hormone receptors. In ten years time, perhaps someone will be writing the closing lines of an introduction to the impact of receptor and antibody chemistry.

REFERENCES

Anon. (1839) *Annal. Pharm.*, **29**, 100.
Fischer, E. (1894) *Chem. Ber.*, **27**, 2985.
Fischer, E. and Slimmer, M. (1903) *Chem. Ber.*, **36**, 2575.
Hoesch, K. (1921) *Chem. Ber.*, **54**, 375.
Robinson, R. (1917) *J. Chem. Soc.*, 762.
Robinson, R. (1974) *Tetrahedron*, **30**, 1477.
Sumner, J. B. (1926) *J. Biol. Chem.*, **69**, 435.
Tigerstedt, R. and Bergmann, P. B. (1898) *Skand. Arch. Physiol.*, **8**, 223.

2 | The mechanistic basis of enzyme catalysis

Ronald Kluger

2.1 INTRODUCTION

The purpose of this chapter is to provide a connection between the concepts of physical organic chemistry and enzyme chemistry through a survey of some of the most useful general approaches that relate the two areas. However, the topic is very large and we have had to be selective. The references we cite in this chapter are therefore intended to lead the reader to more thorough discussions of the points that are raised here. We have selected reviews or extended discussions, rather than citations of experimental observations and original derivations.

Enzymes are noteworthy catalysts. Through the pressures of evolution, they have developed the ability to process the substrates of metabolism with a superb degree of efficiency and specificity which, although remarkable, to a large extent must be understandable within the context of chemical reaction mechanisms since transformations of organic molecules are occurring. It has become a useful pursuit to elucidate the mechanistic basis of enzyme catalysis because, as we understand how enzymes accomplish their tasks, we often discover new catalytic mechanisms. With this knowledge we can begin to design or modify molecules to be useful catalysts and to use enzymes themselves as catalysts outside of their metabolic functions. Seeking a mechanistic basis for enzymic catalysis really means looking for patterns and explanations that we can comprehend and utilize. Thus, it has become apparent that the study of enzyme catalysis is important not only as a traditional biochemical pursuit but also as a means of developing processes of interest for new catalytic systems. The full significance of these remarks will emerge in the next three chapters.

The mechanistic impact of enzyme chemistry has caused an increased interest in the principles of physical organic chemistry. The physical organic chemist is concerned with the properties of organic molecules in systematic relationship to the structure and reactivity of these molecules. Enzyme chemistry is a logical extension of this interest since catalysis and specificity, the central features of enzymic reactions, are concepts that fall within the realm of physical organic chemistry. We shall review some of the areas of physical organic chemistry that relate in a useful way to problems in enzyme chemistry. With this background, we can examine the influence that research into enzyme chemistry has had upon modern approaches to the study of mechanistic and structural organic chemistry.

2.2 CONCEPTS OF CATALYSIS

A survey of the types of reactions catalysed by enzymes reveals that the common categories of organic and inorganic reactions apply. These categories include, for example, substitution, elimination, addition, oxidation and reduction. Such reaction patterns classify the relationship between the starting materials and products; mechanisms provide the means by which a connection occurs. Although the enzyme-catalysed reaction types are obvious, the mechanisms are often conjectural. Where sufficient information is available, it is useful if the reactions can be systematically divided into mechanistic types (such as the Ingold formulation (Ingold, 1953)). The most common examples of this type of classification are the two general nucleophilic substitution mechanisms, S_N1 and S_N2. Walsh (1979) has written an excellent and extensive survey of enzyme-catalysed reactions in terms of reaction type and mechanisms. In addition, several excellent reviews of the mechanisms of enzymic catalysis in general have been published (Bruice and Benkovic, 1966; Fersht, 1977; Scrimgeour, 1977; Metzler, 1977; Jencks, 1969; Bender, 1971; Cunningham, 1978). With these extensive reviews available, I have decided to attempt a somewhat different approach to the subject. I ask the question what are the key concepts of physical organic chemistry that are relevant to considerations of enzyme mechanisms? Illustrative examples, with an admitted bias toward my own areas of interest, will be given.

2.3 DESCRIBING A MECHANISM

A full description of the events that occur during a chemical reaction, especially one that occurs in solution, whether catalysed or uncatalysed, is unattainable because of the large number of independent and dependent variables which describe it. In addition to changes that the substrates undergo during the transformation, one may need to be aware of the function of the solvent and the reorganization that it undergoes (Ritchie, 1969). Obviously, a detailed understanding of any reaction at the level of every molecular co-ordinate is not a practical goal. Yet we would like to know the pathway of a reaction in a mechanistic sense. How does the substrate change during the reaction and what is the molecular function of the enzyme in assuring efficiency and specificity? Can the transformation be quantitatively compared to a reaction which does not involve the enzyme? By having such information we can understand in sufficient detail the function of the enzyme in promoting and controlling the reaction. We will also be able to begin to design catalysts which can be based on what we have learned from analysing the enzymic system (see Chapters 3 and 4).

Kinetic methods are probably the most useful tools for providing a basis on which to analyse the effectiveness of an enzyme as a catalyst. We can usually determine an experimental rate law for an enzyme-catalysed reaction and rate constants associated with the rate law can be evaluated by the use of steady-state methods (Segel, 1975). Thus, a logical sequence of events in terms of conversions between stable species and intermediates can be established.

This general description of an enzymic reaction can be further refined by the use of such techniques as isotope exchange (Segel, 1975), pre-steady-state analysis (Fersht, 1977) and other rapid techniques. With this quantitative set of information, we would like to calibrate the effectiveness of the enzyme as a catalyst against a standard. One useful comparison is with a similar reaction in the absence of enzyme. We know that the enzyme probably catalyses the reaction through a series of steps. It is most helpful if the reaction we are comparing can be set up to coincide step by step with the enzyme reaction. Then, by comparing similar enzymic and non-enzymic kinetics, we can describe the function of the enzyme in terms of specific energetics (see Hall and Knowles (1975) for a good example). In order to do this, we need to be able to analyse the enzymic and non-enzymic systems on a common kinetic basis. The comparison must involve considerable information on both processes, including knowledge of the identity of the rate-determining step.

For these reasons, kinetic methods are especially important and we shall begin by reviewing some of the concepts used by physical organic chemists to simplify the analysis of multistep systems. First, the rate at which reactants are converted to products in any multistep reaction is described by a complex rate equation which can often be simplified using the steady-state assumption (see Hammett, 1970, pp. 77—79). If a reactive intermediate or a reactive complex with a catalyst is on the reaction path, then long before the system reaches its final equilibrium destination, it will reach what is called a steady state. Intermediates present in low concentrations in the steady state undergo *changes* in concentration that are very small compared to those of the observed reactants or products. This method of simplification is particularly appropriate for catalytic systems, since complexes of the substrate and catalyst will usually be present in very low concentrations (Klotz, 1976). In the case of an enzymic reaction, the concentration of enzyme is necessarily low (since its molecular weight is so high). In other reactions, reactive intermediates will also be high in energy and as a result will also be present at low concentrations. A steady state develops as changes in concentrations become undetectable. By this definition, an equilibrium is a steady state that is permanent. However, the steady state we deal with here will be that which is a non-equilibrium condition and will dissipate as components providing the 'fuel' for the steady state is used up. (Life is a steady state; death is an equilibrium.)

2.3.1 Rate-determining step

In order to utilize any rate equation it is necessary to identify a rate-determining step. Murdoch (1981) has presented a particularly clear discussion on this subject. Briefly, in any multistep reaction, steps occurring after the rate-determining step have no effect on the rate of conversion of reactants to products, provided that no intermediate is more stable than the reactant or product. If an intermediate is more stable, then the system should be treated as two separate processes and not as a steady-state system. The rate-determining step is the slowest in rate, not the step with the smallest rate constant. Since rate depends on concentration of reactants and the rate constant, the rate-determining step is a function of concentration of species undergoing that step as well as the barrier to reaction from that step. Steps preceding the rate-determining step provide a flux of material to the state prior to the rate-determining transformation. This means that one can identify a rate-determining step by finding the step with the highest energy-transition state. After a mechanism has been established, we can compare the relative rates of any two processes that an intermediate can undergo. The faster process cannot be rate determining. Comparisons are made for each intermediate until a self-consistent answer is found. For an example in which this method is used to analyse a complex set of data, see Kluger and Chin (1982).

2.3.2 The importance of microscopic reversibility

The principle of microscopic reversibility requires that the lowest energy path in the forward direction is the same as that of the reverse reaction under the same conditions. If there are two competing paths in the forward direction, they will also compete in the reverse direction and to the same extent. Neglecting to take this into account may lead to erroneous conclusions which will give contradictory results. The principle of microscopic reversibility does not require that all molecules react by one pathway — it is a statistical statement that things will distribute themselves according to energy (Hammett, 1970). If the principle can determine with certainty which step is the slowest in one direction, we will know it to be the same for the other direction as well. Competing pathways must compete to the same extent in both directions under the same conditions. We can take advantage of this fact by choosing to examine steps from the direction which is most convenient experimentally. We must confirm what we have learned by studying the reverse process and observing quantitative agreement.

An example of the utility of this principle is the work of Breslow and Wernick (1976) on the mechanism by which carboxypeptidase A catalyses the hydrolysis of peptides. It had been reported by others that the enzyme catalyses the incorporation of oxygen from isotopically labelled water into the terminal carboxyl group of N-benzoylglycine. In any peptide hydrolysis, water must be incorporated into the product carboxylic acid. If the reaction occurs via an acyl-enzyme intermediate, then the enzyme should be able to catalyse exchange of the oxygen atoms of the terminal carboxyl group with those of solvent water in the absence of any free amine (Fig. 2.1). If an amine is required, then exchange does not prove the existence of an acyl-enzyme, since synthesis of a peptide bond also removes oxygen from the carboxylic acid. Subsequent hydrolysis introduces the isotope of the solvent into the carboxyl group (Fig. 2.1). Breslow and Wernick showed that ^{18}O is incorporated from water into the carboxyl group of N-benzoylglycine only when a contaminating amino acid is present, and, in the absence of amino acids, exchange into the carboxyl group does not occur at a kinetically competent rate. Therefore, exchange occurs only by complete hydrolysis and its complete reversal.

2.3.3 The Curtin–Hammett principle

There are few absolute predictive rules in physical organic chemistry but there are many rules that predict trends correctly. One extremely useful and accurate rule is called the Curtin–Hammett principle, first stated specifically by Curtin (1954) who gives credit to Hammett for having taught it to him. The principle is based on transition-state theory. It tells us that the ratio of products of two reactions depends only on their relative rates of formation (Eliel, 1963, pp. 149–156). If the two reactants are conformational isomers, for example, then the barrier to interconversion of the two is usually small relative to the barrier to reaction, and the yield of products from each conformer depends only on the energy of the transition states and not on the energy difference between the conformations. As a result, knowledge of the product ratio tells us the relative transition-state

(1) $RC(O)NHR' + EXH \rightleftharpoons RC(O)XE + H_2NR'$

(2) $RC(O)XE + H_2O^* \rightleftharpoons RC(O)O^*H + EXH$

$RC(O)OH + H_2O^* \xrightarrow{E} RC(O)O^*H + H_2O$

or

$RC(O)OH + NH_2R' \xrightarrow{E} RC(O)NHR' + H_2O$

then

$RC(O)NHR' + H_2O^* \xrightarrow{E} RC(O)O^*H + NH_2R'$

Fig. 2.1 Distinction between two mechanisms of catalysis of carboxyl oxygen exchange catalysed by carboxypeptidase, as reported by Breslow and Wernick (1976).

energy levels but not the relative amounts of starting conformations. On the other hand, if the barrier to interconversion of conformers is larger than the barrier to reaction, then the reaction product ratio does tell you about the conformational population. Product ratios in intermediate cases can be useful if detailed rate information is available.

An elegant application of the principle has recently been presented by Bosnich and co-workers (MacNeil *et al.*, 1981). This work is also related to biomimetic synthesis (Chapter 4). Two intermediates in asymmetric hydrogenation differ as conformers (Fig. 2.2). However, the hydrogenation step from either conformer has a higher energy of activation than that for conformational isomerization. As a result, the transition-state energies and not the conformational populations govern the yield of the product. This fact had not been recognized by previous workers in the area and several mistaken conclusions had been reported.

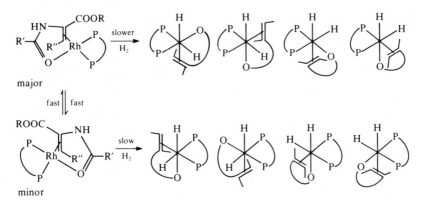

Fig. 2.2 Mechanism of chiral homogeneous hydrogenation, an illustration of application of the Curtin–Hammett principle (MacNeil *et al.*, 1981).

In analysing enzymic reactions therefore, we must only concern ourselves with the effects of substrate conformations if the conformations are interconverted at a rate which is slow compared to the overall transformation; this situation is very rare. Consequently, the advantages of rigid binding of a particular conformation of an intermediate by an enzyme will be small if the conformation is otherwise freely accessible by thermal motions. Interconversions by rotation about single bonds in acyclic structures and most cyclic structures are less than the barrier to reaction. As a result, factors that optimize the conformation of a reactant will also be of little catalytic consequence.

However, conformational effects on enzyme-catalysed reactions can be significant as a basis for reaction specificity. If several products can form from a single intermediate, an enzyme must assure that the product will be the one that is needed and not an uncontrolled distribution. Stereoelectronic control of product selection is a consequence of enzymic control of conformation.

2.3.4   Stereoelectronic effects

The ability of an enzyme to catalyse a single reaction with absolute stereospecificity places very strict limitations on the nature of the catalytic process of that enzyme. How can an enzyme *exclude* reaction paths that would be favourable under non-enzymic reaction conditions from the same intermediate?

Consider the enzyme-catalysed decarboxylation of acetoacetate. The reaction has been shown by Westheimer and his co-workers (Westheimer, 1969) to proceed via an imino-enzyme derived from the amino group of a lysine residue and the carbonyl group of the substrate ((2.1), Fig. 2.3). The

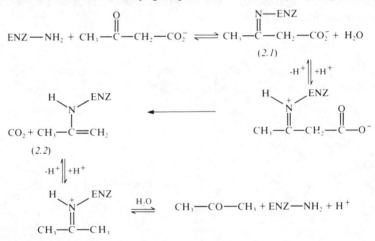

Fig. 2.3 Mechanism of decarboxylation of acetoacetate catalysed by acetoacetate decarboxylase.

product formed after carbon dioxide has been lost is an enamine (2.2) derived from the enzyme and acetone. In order for acetone to leave the active site, the enamine must tautomerize to the imine, which then hydrolyses. Both steps were shown to be enzyme-catalysed since the enzyme catalysed the incorporation of deuterium into acetone from deuterium oxide solvent (Tagaki and Westheimer, 1968). Therefore, a base must be available on the enzyme to remove the proton from the acetone imine to form the enamine and for the reverse process.

These observations appear to suggest that the enzyme might be capable of having dual specificities with a single substrate. If formation of the imine from acetoacetate and the enzyme produces an adduct which will lose carbon dioxide readily, imine formation would normally also increase the acidity of the α-hydrogens derived from the substrate, promoting tautomerization of the imine to an enamine. (The detailed process is described in the next paragraph for a related system.) The enamine would not decarboxylate since conjugative unsaturation is no longer present (2.3). How then does the enzyme prevent this non-productive pathway from being followed?

This problem was originally addressed by Dunathan (1966) with regard to specificity of pyridoxal phosphate-dependent enzymes. The enzymes function by forming iminium derivatives between the amino acid substrate and the aldehyde group of pyridoxal phosphate group of (PLP) (Fig. 2.4). Once the iminium group has formed, the pi system of that group can potentially activate several reactions. For example, protons attached to carbon are normally very strongly bonded to carbon and will not exchange in aqueous base. If, however, the carbon atom to which the proton is attached is adjacent to a carbonyl group or iminium group, exchange occurs fairly readily. The product of removal of the proton is the corresponding enolate or enamine in which there is considerable double bond character between the two carbon atoms with delocalization to the heteroatom. This pi delocalization of the negative charge which would have been localized on carbon in the transition state for proton removal stabilizes the conjugate base of the carbon acid. The transition state leading to this delocalized system must reflect the overlapping character of the orbitals which constitute the pi system of the product. Therefore, the conformation of the transition state must be able to provide for the overlap. Looking at it the other way round, if such a conformation is inaccessible, then labilization of the C−H bond would be prevented.

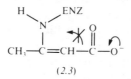

(2.3)

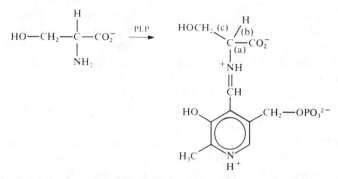

Fig. 2.4 Condensation of pyridoxal 5′-phosphate with an amino acid.

An enzyme can take advantage of this phenomenon very readily. If the substrate is bound in a relatively rigid conformation by electrostatic interactions, then this conformation will control which bonds will be activated when an iminium derivative forms. The enzyme can control the conformation by electrostatic interaction between a positive charge on the protein and the negative charge of the carboxylate of the substrate. In the case of PLP derivatives of amino acids, there are three bonds which could be activated by overlap. The first ((a) in Fig. 2.4) is the C–C bond to the carboxylate; overlap with this bond will cause the enzyme to be a decarboxylase. Overlap of the C–H bond ((b) in Fig. 2.4) will activate the alpha proton; an adjacent base can remove the proton prior to deamination or racemization of the amino acid. Overlap of the pi system and the sigma bond to the side chain ((c) in Fig. 2.4) will permit the enzyme to function as an aldolase.

Convincing evidence has been obtained that PLP-dependent enzymes indeed utilize this feature to control specificity (Dunathan (1967) provides an early review). A number of very specific enzyme inactivators have been designed to take advantage of the mechanism of the PLP-dependent reactions (Abeles and Maycock, 1976; see Chapter 5).

With the concepts of stereoelectronic control of specificity we can solve the dilemma we presented with respect to the specificity of acetoacetate decarboxylase. If the enzyme binds the iminium derivative of acetoacetate in a conformation that prevents the C–H bond from overlapping the C=N of the iminium functional group, then the proton is not activated for transfer to a base (*2.4*). Even if there is a base adjacent to the proton, transfer of the proton will not occur because the transition state cannot be stabilized. However, the bond to the carboxylate group does overlap the pi system, and it will be activated so that decarboxylation can occur without a competitive reaction to slow the overall turnover or destroy the enzyme's specificity.

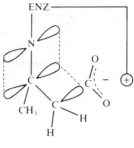

(2.4)

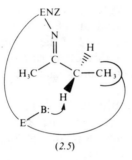

(2.5)

Evidence for this interpretation comes from studies of reactions, catalysed by acetoacetate decarboxylase, of compounds that are not part of its normal catalytic cycle. It was shown that butanone binds to the enzyme's normal substrate-binding site. For example, when butanone was incubated with the enzyme in deuterium oxide, the pro-R proton at C-3 was exchanged much faster than the pro-S proton (Hammons et al., 1975; Benner et al., 1981). (For a discussion of the stereochemical terminology used here, see Eliel, 1980.) The C-1 methyl protons are in homotopic environments and exchange at equal rates. However, since the exchange of the protons of the methylene group involves an enantiotopic selection, that reaction must be occurring in the chiral environment of the active site (2.5). The exchange is much faster than in the absence of enzyme, so the enzyme must be actively catalysing the exchange process, presumably via formation of the iminium derivative. For a further example, see Kluger and Nakaoka (1974).

All enzymic reactions that involve breaking or formation of C–H or C–C bonds should be subject to similar analysis. Recently we analysed the most likely structures of adducts of thiamin and pyruvate, on the basis of extrapolations from the X-ray crystal structure of the phosphonate analogue of

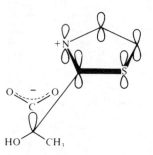

Fig. 2.5 Adduct of thiamin and pyruvate. In this conformation, either CO_2 or pyruvate can be eliminated (from Kluger, 1982). Reprinted with permission of copyright holder (New York Academy of Sciences).

the adduct (Turano *et al.*, 1982). The structure that results from the maximal overlap direction in formation of the C–C bond between pyruvate and thiamin also aligns the C–C bond of the carboxylate group so that it will overlap the pi system of the thiazolium ring of thiamin ideally placed for decarboxylation. Thus the two main catalytic steps are structurally correlated (Fig. 2.5).

In summary, although conformational factors in the substrate have little effect on accelerating any step in enzymic catalysis, they are often crucial in controlling selectivity. We must now consider the origins of catalysis.

2.4 ENTROPY AND ENZYMIC CATALYSIS

2.4.1 *The molecular basis of entropy*

Another application of physical organic chemistry to enzyme catalysis is the analysis of the means by which enzymes accelerate reactions of bound substrates. Most current discussions on the subject include a consideration of the relationship of the loss of entropy on binding of the substrate to the enzyme to gains in catalytic efficiency (see Jencks (1975) for a thorough discussion). Detailed treatments are presented which relate entropy effects to catalytic rate constants and later in this chapter we will present such a discussion. In order to make this more useful, however, it is probably a good idea at this point to review the general concepts and equations related to the relevant aspects of entropy.

Thermodynamics and statistical mechanics arrive at expressions for free energy in terms of enthalpy and entropy. Since free energy is related directly to chemical equilibria, it is the thermodynamic state function of most use in studying chemical phenomena. The change in free energy with temperature at constant pressure is defined as entropy. The definition of Gibbs free energy (G) is given by:

$$G = H - TS$$

where H is enthalpy, T the absolute temperature and S entropy. Thus the derivative of free energy with respect to T is S. The relationship of this type of equation to molecular phenomena and mechanisms comes from applications of statistical mechanics (Denbigh, 1964, part III):

$$G = E - RT\ln Q + RT$$

The quantity Q is a partition function which is the sum of probabilities of molecular energy states; E is thermal energy; and R is the gas constant. Solving for G in the two expressions for free energy given above yields a single equation which can be used for defining entropy:

$$S = (E/T) + R\ln Q$$

Since E/T is held constant, all that is needed in order to find a numeric value for entropy change is the difference in the $R\ln Q$ term. To go from A to B the entropy change will be $R\ln(Q_A/Q_B)$. If the partition functions for the two species can be derived, so can the contribution of entropy to the change in free energy.

The partition function Q can be factored into the components that sum to the total energy of a chemical species: its translational, vibrational and rotational energy contributions at constant electronic energy. The translational contribution can be calculated by using a statistical analysis of the number of ways particles can be distributed among energy levels. A particle of mass m in a container of volume V has a partition function:

$$Q = (2\pi mkT)^{3/2}V/h^3$$

where h is Planck's constant.

Since the arrangement of species and energy levels is independent of the identity of the particles, the expression for translational entropy of N molecules is statistically corrected:

$$S = R[5/2 + \ln(2\pi mkT)^{3/2}V/Nh^3]$$

This expression is known as the Sackur–Tetrode equation and its validity in providing accurate values for translational entropy of atoms and molecules has been confirmed experimentally in a large number of cases over the last 50 years. Thus we can use this expression to estimate the translational entropy of a molecule before it comes in contact with an enzyme.

The Sackur–Tetrode equation predicts that a species such as a substrate of molecular weight 100 in solution at a concentration of 10^{-6} M will possess about 60 entropy units of translational entropy. At 300 K, therefore, the

immobilization of the molecule by its becoming bound to an enzyme leads to
the loss of translational entropy corresponding to a deficit in free energy of
$18 \, \text{kcal} \, \text{mol}^{-1}$ ($75 \, \text{kJ} \, \text{mol}^{-1}$). This must be compensated for to a significant
extent by the favourable interactions which lead the substrate to associate
with the enzyme.

Rotational and vibrational entropy can also be calculated from knowledge
of the appropriate partition functions. Rotational contributions to the overall entropy of a molecule are considerably smaller than translational contributions. The binding of a substrate to an enzyme therefore does not usually
result in a significant change in this quantity. Vibrational contributions are
usually small enough to be ignored.

In summary, the translational entropy of a molecule can be approximated
from the Sackur–Tetrode equation on the basis of the derivation presented
above in which the statistical and classical definitions of free energy are
equated.

2.4.2   Binding and enzyme catalysis

It is obvious that the chances of a bimolecular reaction occurring between
species in dilute solution is less than it is if those two species are adjacent to
one another. In order for them to be adjacent for a significant interval of
time they must be held there by a third party which assumes the role of
catalyst. One very important role for an enzyme is to function in this way. It
must be able to encourage two species to react by ensuring that they are in
proximity to one another and that they are aligned in such a manner that
their contact can lead to a reaction.

The substrate-binding sites of enzymes possess functional groups at which
it is energetically favourable for the substrates to associate. A considerable
loss of entropy occurs when a molecule which has been free to move about in
solution becomes associated in a specific manner with an enzyme. This is
compensated for by favourable binding interactions (Westheimer, 1956).
This form of enthalpic 'trapping out' of entropy by an enzyme applies to any
number of different types of substrate molecules which interact with the
enzyme and which eventually must interact with one another. If the entropy
loss on binding has been compensated for by the favourable enthalpic interactions which also result, then the molecules associated with the enzyme
have undergone a productive transformation from randomly moving indi-
viduals in a solution to productively aligned reacting partners on a catalytic
surface. They can be considered to be appendages of the enzyme surface and
as such two reacting species have become parts of the same molecule. If they
are held in a manner that encourages reaction, then a considerable part of the
entropic barrier to reaction has been removed (Jencks, 1975).

How to establish whether, and to what extent, a situation favourable to reaction will be created in any particular case was the subject of considerable controversy in the 1960s and 1970s. Several approaches were popular and then went out of favour as inaccuracies were discovered. Although the question may not be fully settled yet, it appears that a general consensus has been reached that a logical and justifiable analysis can be made by considering the entropic consequences of binding by comparing the situation to that of an intramolecular reaction as explained in the following paragraphs. A complete treatment of this topic is given by Jencks (1975) and Bruice (1970).

2.4.3   The intramolecular analogy to enzyme catalysis

In a bimolecular reaction, two molecules possessing translational entropy lose a considerable amount of that entropy when they interact. Since the two species must not separate from one another if they are to react, for the purposes of entropy analysis the two molecules can be treated as a single entity. As such, the equivalent of the translational entropy of one of the two is lost. In considering the reduction in translational entropy caused by binding of a substrate to an enzyme, we saw that we can use the Sackur−Tetrode equation to arrive at a numerical value for the entropic change. A similar change should take place when a bimolecular reaction occurs.

In an enzymic reaction the entropic loss caused by binding is compensated for by the favourable enthalpic interactions which occur when the substrate associates with the enzyme. For the reaction between two free species in solution, in the general case, no significant compensating enthalpic interactions occur. Of course, molecules exist which can have favourable interactions with their reaction partners and sometimes these are considered to be 'model enzymes'. This topic will be covered in the section of this review on biomimetic chemistry and, in the context of synthesis, in Chapter 4.

In an intramolecular reaction, the immobilization of a reactant that is present at a concentration of 1 M corresponds to a loss in translational and rotational entropy of about 35 entropy units (10.5 kcal mol^{-1} (44 kJ mol^{-1} at 300 K)). If the species is previously immobilized by being covalently associated with its reaction partner in a rigid manner, then this entropy loss can be avoided. This corresponds to an increase in an equilibrium constant of about 4×10^7.

The most common way of calibrating an intramolecular reaction against its bimolecular counterpart is through the quantity called *effective molarity* (see Kirby (1980) for a thorough discussion of this topic). A bimolecular reaction is characterized by a rate constant for a reaction that obeys second-order kinetics. The rate constant k has units M^{-1} s^{-1} but the corresponding intramolecular reaction has units of s^{-1}. The ratio of the rate constant for the intramolecular process to that for the bimolecular process will not be a dimensionless number but will have units of molarity. The particular intramolecular analogue of the bimolecular reaction being studied will be characterized by this ratio, the effective molarity, with units of molarity (Kirby, 1980). Since the bimolecular reaction can be brought to comparable units by multiplication by molarity, the intramolecular reaction can be said to be accelerated relative to the bimolecular reaction by an apparent concentration factor. The meaning of this factor in terms of physical reality has been controversial. Values over 10^8 M have been found for the effective molarities of intramolecular reactions, yet this cannot correspond to a physically attainable concentration. Does this then mean that the intramolecular phenomenon must be considered from a basis other than this apparent concentration?

The question of the basis of the intramolecular effect has been addressed in detail by Page and Jencks (1971) and this discussion is based primarily on their conclusions. Since effective molarity is really only the ratio of rate constants, it need not correspond to a phenomenon that is achievable only by increasing concentration; other factors are obviously involved. Two reacting species which are held together have the translational entropy of a single species and roughly the rotational entropy of the two separate species. Two separate reactants each possess the full complement of translational entropy. Combining two reacting species into a single molecule eliminates the translational entropy of one molecule. Since this translational entropy will have to be lost anyway in order for the bimolecular process to occur, what actually happens in the intramolecular process is a pre-activation towards the transition state for the bimolecular reaction. This measures the entropic advantage of the intramolecular process in units of molarity. Examples of reactions with high effective molarities will be presented in Section 2.7.

2.5 PROTON TRANSFER AND CATALYSIS

As we have said, high effective molarities result when reactions involve the coming together of two reacting molecules and values of over 10^8 M can be observed. Reactions carried out in the laboratory are often subject to catalysis by acids or bases but reactions involving intramolecular general acid catalysis or general base catalysis never show effective molarities over about 60 M. Bernasconi and co-workers (Bernasconi et al., 1982) have provided an incisive analysis of the basis of effective internal proton transfers. Factors which control the effectiveness of the transfer process include: ring size for the internal proton transfer (six-membered ring is most effective), acidity and basicity of the donor and acceptor sites, acidity of the reaction medium, and the sensitivity of the proton-transfer rate to the basicity of the acceptor. An important conclusion with respect to enzymic catalysis is that large rate accelerations do not occur by facilitation of single proton-transfer steps. However, in combination with other steps, the factor is multiplicative. An ineffective proton transfer will have a major retarding effect that is also multiplicative.

The transfer of a proton between a Bronsted acid and a Bronsted base occurs as part of most enzyme-catalysed reactions. The details of proton-transfer reactions have been worked out and volumes devoted to the subject have appeared (for example, Bell, 1973). In the case of enzymic reactions, proton transfer is usually coupled to some other bond-forming or bond-breaking process.

Eigen and his co-workers established the fundamental properties of proton-transfer reactions (Eigen, 1964). Transfer of a proton in the thermo-dynamically favoured direction occurs at a rate which has diffusion of the proton in water as its rate-limiting step. Proton transfers that occur at the diffusional limit therefore will have similar rate constants. Eigen determined that these rate constants are of the order of 10^{10} per second. If the equilibrium constant for the proton-transfer reaction is known (from the ratio of acid dissociation constants of the two partners in the reaction), then the rate constant for proton transfer in the 'slow' direction can be calculated. If the reaction is symmetric in energy, then neither direction will be at the diffusional limit and both the forward and reverse reaction rates will be comparable. A plot of the rate constant for transfer of a proton from an acid to a series of conjugate bases is called an Eigen plot. A typical plot is presented (Fig. 2.6).

For reactions in which proton transfer occurs in conjunction with some other bond-making or -breaking process (a concerted system), the analysis becomes more complex. How can we be sure that the two steps are really occurring together and not in rapid succession? (For a review see Jencks (1980).) The problem is a significant one because it sets the limits to the efficiency of acid–base catalysis.

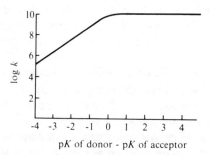

Fig. 2.6 Second-order rate constant ($M^{-1} s^{-1}$) for transfer of a proton from an oxygen or nitrogen acid to a series of oxygen or nitrogen bases. The phrase 'pK acceptor' refers to the conjugate acid of the acceptor. The rate constant reaches a maximum value which is that for diffusion of a proton in water.

A reaction that is accelerated by the presence of Bronsted acids other than protons is said to be general-acid-catalysed. This is easily diagnosed since addition of a Bronsted acid (i.e. a buffer) at constant acidity will increase the rate of the reaction. Since a buffer consists of an acid and base component, the effect of changing the buffer ratio must be assessed to determine whether the acid or the base component is catalytic. Occasionally both components are active. For general acid catalysis, the stronger the acid catalyst, the larger should be the observed rate constant. A plot of the logarithm of the observed rate constant versus the pK_a of the catalytic acid will normally be a straight line provided that the mechanism and rate-determining step remain constant. This plot, called a Bonsted plot, is an extremely useful tool for analysing mechanisms related to enzymic catalysis.

The catalysis of a reaction by buffers suggests that transfer of a proton is involved in the rate-determining step, but it does not identify that step. The rate law that is obtained describes kinetically equivalent mechanisms (see Bell, 1973). Furthermore, is proton transfer occurring alone or in conjunction with some other process? Jencks (1980) has argued that proton transfer may occur in a single step along with another bond-breaking or -forming process if that is a lower-energy pathway than one in which the processes occur in separate steps. These rules must apply to enzyme-catalysed reactions as well since proton transfers often occur in conjunction with other processes. A detailed analysis of enzyme reactions at a similar level of detail is becoming possible through the interpretation of phenomena such as solvent isotope effects (see Schowen, 1977).

2.6 ENZYMIC EFFICIENCY

Let us consider an enzyme that catalyses a multistep process which converts a single substrate to a single product. Just how 'good' is this particular enzyme as a catalyst compared to other proteins which might catalyse the same reaction? It has been argued convincingly by Hanson and Rose (1975) that enzymic efficiency is an evolutionary process subject to mutation and improvement. We can see how to set a standard if we consider the basic facts of the system.

The thermodynamic relation of reactants and products is independent of the catalytic process. Thermodynamic analysis of the reaction tells us whether the conversion will result in the net gain or loss of chemical energy to the system, which will be converted to heat and/or work. An endergonic process will require an energy input and will always have a barrier at least equal to the thermodynamic difference between the starting material and product. An exergonic process will not have a thermodynamic barrier to overcome but it will be limited by the rate of diffusion of the substrates to and of the products away from the enzyme. The endergonic process has this diffusion barrier superimposed upon the thermodynamic barrier. To simplify our analysis of efficiency, we shall eliminate consideration of the thermodynamic barrier but will remember that it must be superimposed upon any conclusion we reach.

We know that in any multistep reaction, the most significant barrier is that of the rate-determining step. If that barrier is lowered, the reaction will proceed faster. (The rate expression is always reducible to the rate constant for the rate-determining step times the concentration of the species reacting in the forward direction in that step adjusted for the extent of reverse reaction.) An enzyme can become more efficient if it can lower the barrier of the rate-determining step (by such devices as more reactive functional groups or better relative placements of binding sites, for example). However, there must be a limit to this since, as the barrier for that step becomes sufficiently low, another step will then have a higher absolute barrier and thus will become rate-determining. This step will then be subject to evolutionary pressure, and it will improve until another step takes over. The limit of this improvement is reached when all steps have transition states which are comparable in energy (corrected for thermodynamic barriers). That is, each is partially rate-determining (for an example see O'Leary and Baughn, 1972). The lowest absolute barrier that can exist for any reaction in solution is that for diffusion; therefore, the ultimately most efficient, isolated enzyme will be that which catalyses all steps with barriers equal to that for diffusion.

Albery and Knowles (1977) developed a mathematical function that describes the efficiency of an enzyme as a catalyst based on the assumption that the ultimate rate-controlling factor for any single enzyme-catalysed reaction must be the rate of diffusion. (Multistep metabolic processes have more complex constraints since each reaction is dependent upon the efficiency of other reactions in the pathway.) If all chemical processes in enzyme mediated reactions are subject to highly effective catalysis, then the overall reaction will be limited by diffusion. The quantitative application of this concept requires detailed mathematical information that is given in extensive articles which are cited in the review presented by Albery and Knowles (1977). A cursory survey reveals that the concept has not been widely utilized as yet, probably because sufficient details about individual steps of enzymic reactions are not known. In the future, the provision of such detailed information should become an important goal for research.

Substrates undergoing enzyme-catalysed reactions must be able to arrive and depart from the active site of the enzyme without difficulty. We have considered that the rate of diffusion of substrates and products ultimately determines the productivity of catalysis. If a conversion is to occur, the reactants must be present at the active site simultaneously. The products must also be able to escape readily after a reaction has occurred. An enzyme with a very high affinity for the reaction products would cause the reaction to 'back up' toward the transition state of the step prior to the diffusion step. If this barrier is lower than that for diffusion, the efficiency of the catalytic process is compromised.

This may be a serious problem in catalysis for very efficient enzymes or those that deal with large substrates which can interact strongly at many sites, for example, peptidases and nucleases. One way an enzyme can turn the affinity problem into an asset is to use the binding energy in a productive way to achieve catalysis. A tightly bound product will be released at a rate less than the diffusion-controlled maximum. However, if the energy that is available through the interaction of the product and the enzyme could be used to cause the enzyme to isomerize to a form that has a lower affinity for the product, then catalysis would be facilitated. However, once the product has adhered to the form of the enzyme for which it has an affinity, it locks out the chance for a change. The isomerization should take place before the last step has occurred to avoid the product blockage. The formation of covalent intermediates in enzymic reactions may help to overcome this problem.

Consider the case of pepsin, an enzyme that catalyses the hydrolysis of peptide bonds in large substrates at an early stage of digestion in the stomach. Evidence has been accumulated that suggests that the reaction may occur by the formation of sequential covalent intermediates between the enzyme and the two portions of the peptide bond (see the discussion in Kluger and Chin (1982) and Fersht (1977)). Since the substrates are large, they have a high affinity for the enzyme and diffusion of the products away must be a problem. If sequential covalent intermediates form, then the formation of the intermediate provides the energy that enables an isomerization of the enzyme to occur. The isomerized enzyme can then release the non-covalently held substrate readily.

2.7 EXAMPLES OF INTRAMOLECULAR CATALYSIS

One of the most effective procedures for gaining information about the nature of interactions between the functional group of a substrate and a catalytic group on an enzyme is the use of intramolecular model reactions. Our general analysis of the basic characteristics of enzymic catalysis suggests that entropic effects play a major role in an enzyme's processing of its substrates. The prior immobilization of interacting functional groups relative to one another enables them to react without first overcoming the barrier that is due to losses of translational entropy in bimolecular reactions. Many interactions which do not have very high enthalpic barriers are masked by the built-in entropic barrier. Thus one may falsely conclude, on the basis of the low reactivity of bimolecular models, that a particular substrate is unreactive towards the functional groups at the active site of an enzyme.

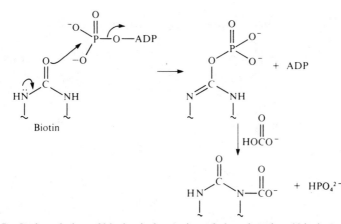

Fig. 2.7 Carboxylation of biotin via formation of phosphorylated biotin (see Kluger *et al.*, 1979).

For example, one likely mechanism for the ATP-dependent carboxylation of biotin involves nucleophilic attack of the ureido group of biotin upon the terminal phosphate of ATP to produce *O*-phosphobiotin (Kluger and Adawadkar, 1976). The reactivity of urea towards the phosphate functionality was unknown because any bimolecular reactions between those functional groups had been too slow to observe. However, this can be overcome by immobilizing the groups on a single molecule so that no loss of translational entropy occurs when the two functional groups interact. We found that the urea reacted readily with the adjacent phosphate derivative. The entropic barrier to bimolecular reactions had obscured the likely interaction of these two functional groups (Kluger *et al.*, 1979). The mechanism has thus been established as a serious possibility for the enzymic carboxylation of biotin (Wood and Barden, 1977) (Fig. 2.7).

Carboxyl derivatives also exhibit strong neighbouring group effects. For example, an amide that can react with a neighbouring carboxyl is hydrolysed much more rapidly than an isolated amide (Bender, 1957). The amide reacts first with the carboxyl group to form an anhydride with release of the amine portion of the amide (Fig. 2.8). The anhydride then reacts with water to produce the carboxyl group derived from the amide and also to regenerate the carboxyl group that was originally present.

The first compounds of this class that were subjected to mechanistic study were derivatives of phthalamic acid. Bender and his co-workers used isotope labels and kinetic analysis. They found that catalysis of the hydrolysis of the amide occurs by intramolecular formation of an acylated orthoamide by addition of the carboxyl group to the neighbouring amide (Bender, 1957; Bender *et al.*, 1958). The form of the molecule in which the carboxylic group is undissociated, or some tautomeric equivalent, is reactive whereas the

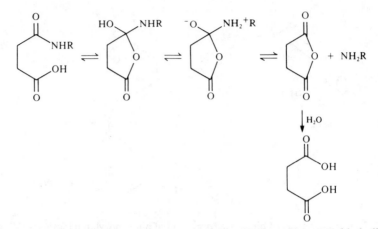

Fig. 2.8 Participation of an adjacent carboxylic acid at an adjacent amide facilitates hydrolysis by formation of an anhydride.

conjugate base is stable. It was proposed that the carboxyl group adds to the amide group in a four-centre reaction to produce the cyclic orthoamide intermediate (Bender *et al.*, 1958). This intermediate decomposes to anhydride and amine. The key to the efficiency of this route is the ease of formation of the orthoamide intermediate.

Kirby and his co-workers studied the reaction in greater detail (Kirby and Lancaster, 1972; Aldersley *et al.*, 1972). Through a series of rigid substituted derivatives of maleamic acid, they demonstrated that ground-state strain dramatically increases the rate by promoting cyclization, giving effective molarities of up to 10^{10}. They also showed that formation of the intermediate is a faster step than is its decomposition to products. In other words, the conversion of amic acid (1) to anhydride and amine involves rate-determining breakdown of the intermediate.

2.8 TRANSITION-STATE ANALOGUES

At the beginning of this chapter, we argued that the organic chemical concept of reaction mechanism should be logically applied to enzyme reactions, but we have also seen that the problem is even more complex than in the usual organic reaction because the important and large enzyme molecule must be accounted for. Since the enzyme is much larger than any usual substrate it provides the environment of the reaction. In a complete analysis, accounting for the enzyme's course during the reaction is as important as accounting for the transformations of the substrate. However, it is much easier to follow structural changes of the substrate than it is to follow the conformational changes of the enzyme. Since the enzyme is a catalyst, its structure before and after a reaction has occurred should be constant. The substrate, on the other hand, has undergone a series of changes which have converted it into the product. Examination of the enzyme before and after the reaction obviously gives no information. Observing the enzyme directly during catalysis is important and possible but usually difficult. One way of simplifying this task is to monitor the enzyme by using a substrate that reports the nature of its interactions with the enzyme. Spectroscopic observation of a substrate or of a modified substrate often gives extremely useful information in this connection. (For a recent example of a modern spectroscopic approach see Huber *et al.*, 1982.) An alternative approach is to prepare inhibitors which are substrate variants designed to interact with a transient but catalytically significant form of the enzyme. The effects of the inhibitor can then be examined kinetically, spectroscopically, or, as recently reported, by product analysis (MacInnes *et al.*, 1982). Let us consider the various steps involved in the catalytic process to see where the approach can be used.

Enzyme-catalysed reactions involving a single substrate which is converted to a single product present a basis for analysis of reaction mechanisms. Multiple substrate reactions add problems that must be dealt with from the basis established for unimolecular processes. For a unimolecular conversion, we know that the substrate must bind to the enzyme, be converted to the product during the catalytic process and then separate from the enzyme.

With the substrate bound at the active site of the enzyme we can assume, for simplicity, that the loss of entropy due to binding is more than compensated for by favourable interactions between the substrate and the enzyme. This initial state must be such that the substrate is relatively unaltered from its structure before binding. However, to undergo reaction the substrate must proceed to some higher energy form, toward the structure of the transition state. The enzyme must encourage this process. Pauling originally proposed that an enzyme could catalyse a reaction by assuming a conformation that would stabilize the transition state particularly well (see Wolfenden (1972) for a discussion of this point). If the enzyme can alter its conformation to a form with a high affinity for the transition state, then the reaction will be catalysed if the price of the isomerization in terms of energy is less than the amount of transition-state stabilization that will be achieved.

Since the transition state occurs at the highest energy point on the reaction surface, it is necessarily an unstable arrangement and no method will be available to isolate it. Yet, if the goal of the enzyme is to stabilize this species, the enzyme should be complementary in form to the transition state whose formation it is promoting. If we could isolate the form of the enzyme that stabilizes the transition state, then we would have a 'negative image' of the transition state. This means of analysis was proposed by Wolfenden (1972) and by Lienhard (1973). They also recognized that the enzyme will exist as a collection of forms that bind substrate and transition state. The formation of the enzyme—substrate complex can provide the energy necessary to isomerize the enzyme to the form that will stabilize the transition state. (Jencks (1975) has called this the Circe effect and has written an extensive analytical review on the subject.)

Experimental evidence has also suggested that species that bind much more tightly to enzymes than the natural substrate might be doing so because they resemble the transition state for the reaction (Cardinale and Abeles, 1968). For example, the transition state for the interconversion of D- and L-proline, catalysed by the enzyme proline racemase, should resemble the planar species, pyrrolecarboxylic acid (Fig. 2.9). The observation that pyrrolecarboxylic acid binds more tightly to the enzyme than does either substrate is consistent with its being a structural analogue of the transition state for the interconversion catalysed by the enzyme (Jencks, 1969, pp. 300–301). Wolfenden (1972) and Lienhard (1973) realized that in addition to this being a useful *ad hoc* explanation of a single phenomenon, it is also a basis for a general approach to analysing transition-state structure and to inhibitor and drug design (see Chapter 5 and Wolfenden, 1979).

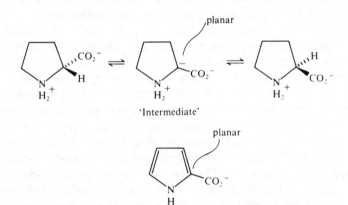

Fig. 2.9 Enzyme-catalysed interconversion of D- and L-proline proceeds via a planar transition state which can be mimicked by pyrrole-2-carboxylate.

This approach has been very useful for designing inhibitors of very specific enzymes and numerous examples are cited in review articles. However, we must be careful to realize that we are undertaking both the experiment and interpretation with considerable prejudice. We assume that tight binding is occurring because a molecule resembles the expected transition state for the reaction. Even if we are right and can exclude other transition states from contention, we must still make sure that we are really dealing

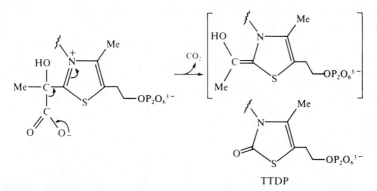

Fig. 2.10 Conversion of pyruvate adduct of TDP involves a transition state that probably has character associated with neutral enamine formed as CO_2 is lost. TTDP is an analogue of the enamine.

with a proper cause and effect relationship; we must exclude all other possible reasons why the analogue and the enzyme would have a high affinity for one another. However, since we cannot know what all other possible reasons are, we are left with experimental evidence that at its best can only be suggestive and never conclusive.

For a mechanistic example, let us consider one of the classic cases of a successfully designed transition state analogue. Pyruvate dehydrogenase catalyses the oxidative decarboxylation of pyruvate. The single substrate for decarboxylation is pyruvate which forms a covalent adduct with thiamin diphosphate (TDP) (Kluger and Smyth, 1981) at the active site of the enzyme. This reaction is also discussed in relation to biomimetic synthesis in the next chapter. The transition state for the decarboxylation reaction involves conversion of a zwitterion to a species that has much less charge separation. The transition state therefore will be best reflected in a molecule whose structure does not contain the zwitterionic features of the starting adduct. Gutowski and Lienhard (1976) reasoned that thiamin thiazalone diphosphate (TTDP) resembles the transition state for the decarboxylation process in terms of its charge distribution (Fig. 2.10).

In accord with the prediction, TTDP was found to bind very tightly to pyruvate dehydrogenase, binding much more tightly than do thiamin diphosphate or pyruvate, perhaps because it is a transition-state analogue. However, as we argued above, there may be other unanticipated factors that cause it to bind tightly. For example, the TDP-binding site of this enzyme and other TDP-dependent enzymes appears to be low in polarity, so that analogues of TDP that are not zwitterions in general bind well to the TDP-binding sites. For example, thiochrome diphosphate (*2.6*) is a tricyclic planar molecule that is not a close structural analogue of TDP or of a transition state, yet it binds well to the TDP-binding site of pyruvate decarboxylase (Wittorf and Gubler, 1970).

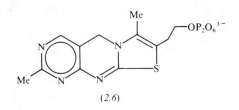

(2.6)

It is also difficult to tell a transition-state from a reactive intermediate, since they may be close in energy and structure (the Hammond postulate). Thus, the phrase 'transition-state analogue' is often replaced with the more conservative 'reactive intermediate analogue' (Byers, 1978). However, since the two types of species should resemble one another, the difference is not serious, for the purpose of analysis. Wolfenden (1974) has suggested that a transition state analogue may not bind to an enzyme if the enzyme isomerizes as a result of forming the transition state. The substrate must bind and then both enzyme and substrate change together to the transition state. Enzyme alone does not isomerize so that the form that would recognize the transition state analogue is not present in the absence of substrate.

As long as these limitations to the method are considered, transition-state analogues can be very useful tools in elucidating enzymic mechanisms and in the design of inhibitors. By observing an enzyme–analogue complex spectroscopically, it is potentially possible to observe the structure of the enzyme form that is specifically responsible for catalysis.

2.9 MULTIPLE BINDING SITES

Reaction sites on substrates in enzyme-catalysed processes are often small parts of larger molecules. The affinity of an enzyme for a small molecule is limited, since the attractive interactions between enzyme and substrate depend on the sum of contributions of each structural component of the small molecule with the enzyme (Jencks, 1975). The larger and more complex the substrate molecule, the more possible sites for favourable energetic interactions with the enzyme are likely to exist. The interactions of portions of the substrate with the enzyme are primarily enthalpic and include hydrogen-bonding, electrostatic attractions, dipolar alignments and apolar solvent exclusion.

The organic coenzymes provide good examples of molecules with many sites that can interact favourably with enzymes (Bruice and Benkovic (1966), volume 2). It is likely that their evolutionary success has been partially due to their multiple sites for interacting with the protein with which they associate. For example, the nicotinamide adenine dinucleotides contain phosphate groups that can make strongly favourable electrostatic interactions with cationic sites on the enzyme to which they bind. In addition, the ribosyl moieties of the coenzyme contain hydroxyl groups that can hydrogen-bond to acceptor sites. Similar nucleotide-like units in other coenzymes, such as FAD and thiamin diphosphate, can also assist in binding (Metzler (1977), Chapter 8).

The binding interactions of substrates and enzyme-bound thiamin diphosphate are illustrative. Thiamin diphosphate binds non-covalently to the apoenzyme with which it is associated with a relatively high affinity (Krampitz, 1970). In the case of pyruvate decarboxylase, the substrate pyruvate and the product acetaldehyde are small molecules and have little to offer for favourable enthalpic interaction with the catalytic site of the enzyme. During the catalytic cycle of the enzyme, thiamin diphosphate and pyruvate combine to form an adduct, lactylthiamin diphosphate, which exchanges carbon dioxide for a proton, yielding hydroxyethylthiamin diphosphate (Kluger, 1982) (Fig. 2.14). This product decomposes to acetaldehyde and thiamin diphosphate (Krampitz et al., 1961). The thiamin diphosphate molecule remains associated with the apoenzyme while acetaldehyde is released into solution. During the course of catalysis, the small, weakly bound pyruvate molecule is converted into a fragment of the larger lactylthiamin diphosphate molecule. The formation of this adduct between thiamin diphosphate and pyruvate is thus the key step of the catalytic cycle.

The importance of binding in catalysis has been emphasized in this chapter and we might expect that formation of lactylthiamin diphosphate would be a process that increases the affinity of the enzyme and pyruvate. However, evidence has now accumulated that the adduct does not improve the affinity of pyruvate for the enzyme (Kluger and Smyth, 1981). In fact, it has a markedly lower affinity than does thiamin diphosphate itself. Where, then, is the catalytic advantage of adduct formation? The adduct provides a route to the transition state for the decarboxylation process, which, as we already have shown is more like a neutral thiazole (Breslow, 1962; Gutowski and Lienhard, 1976). Clearly, if the enzyme has a high affinity for a neutral thiazole it cannot also have a high affinity for zwitterionic lactylthiamin. Experimental evidence for this remarkable situation was obtained as follows.

The affinity of the apoenzyme of wheatgerm pyruvate decarboxylase for the lactylthiamin diphosphate adduct must be very low (Kluger and Smyth, 1981) because we could detect no activation of the enzyme by the adduct above the level of a control. However, the t-butyl ester of lactylthiamin diphosphate binds and activates the enzyme with a K_m about twice that of thiamin diphosphate. Thus, these affinity measurements do not reflect a steric sensitivity to the bulk of the pyruvate group. If thiamin diphosphate serves as a high-affinity 'anchor' (Jencks, 1975), then a decrease in affinity of lactylthiamin diphosphate relative to the unsubstituted species indicates that the lactyl group has a negative affinity for the apoenzyme. Since the ester binds, this indicates that the negative effect is due to the presence of an unesterified carboxyl group on lactylthiamin diphosphate.

2.10 BIOMIMETIC CHEMISTRY

A logical extension of the acquisition of principles that we have discussed so far is the application of these principles to new situations. If we can explain how an enzyme works in terms of chemistry, then we should be able to design other chemical catalysts whose properties are predictable from what we know about enzymes. Breslow coined the expression 'biomimetic chemistry' to deal with those aspects of the field whose goal is to design and build molecules whose properties rival enzymes as catalysts. (For a recent review see Tabushi (1982).) It has become a very popular activity to design molecules which have predictable enzyme-like characteristics, that is, they are efficient and specific catalysts which have a high affinity for their substrates. We will illustrate some of the progress in the design of catalysts aimed at promoting catalytic efficiency. Chapter 4 deals with catalysts designed for synthetic utility.

A major effort of physical organic chemists who pursue research in bio-mimetic chemistry is the production of 'model enzymes'. Their goal is to produce a true, efficient and specific catalyst by rational design. Usually, the approach that has been taken has involved the conversion of a naturally occurring species that has some ability to bind the desired substrate. For example, a series of oligosaccharides that are cyclic $(1-4\alpha)$ hexamers of glucose, known as Schardinger dextrins, can be obtained by microbial degradation of starch. Variants containing six glucosyl units (cyclohexamyloses), seven units (cycloheptamyloses) and eight units (cyclo-octamyloses) can be obtained. They are generally called cyclodextrins when no ring size is specified. The structures of these species have been carefully studied and it appears that in general they exist as toroids (doughnut-shaped solids). The hydrophilic hydroxyl groups of the glucose units are directed toward the solution while C−H and C−O−C groups are directed toward the inner cavity (2.7) Apolar molecules have an affinity for the inner cavity of the cyclodextrin and will form stable complexes. Considerable effort has gone into making catalysts based on the cyclodextrin molecules (Tabushi, 1982).

Early examples of catalysis by cyclodextrins usually involved reaction of a hydroxyl group of the cyclodextrin with an unsaturated electrophilic centre on the substrate. Thus Bender found that the acylation of a cyclodextrin by m, t-butylphenyl acetate occurs with a rate constant that is 250 times larger

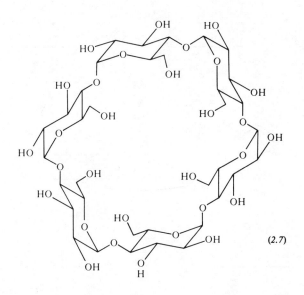

(2.7)

than the rate constant for hydrolysis (van Etten *et al.*, 1967) (Fig. 2.11(a)). Although this is a significant acceleration, it is not typical of accelerations of $k_{cat.}$ expected for an enzyme-catalysed process. Breslow and his co-workers (Breslow *et al.*, 1980) carefully analysed the structural features of the cyclodextrin and the substrate to try to design improvements into the catalytic system. They found that the reaction Bender had studied is slowed down because, although the substrate is bound effectively within the cyclodextrin cavity, the hydrolysis intermediate tends to move out of the cavity. To prevent the intermediate from leaving the cavity, the cyclodextrin was modified by having a synthetic capping moiety placed across one side of the cavity. Further, substrates were found that fit the cavity better and thus would have a smaller tendency to leave. After examination of molecular models, it was concluded that ferrocene should bind especially well to cyclodextrins so that esters derived from ferrocene should make excellent substrates. This led to increases in the catalytic rate constant of 1000-fold over those that had been observed by Bender and his co-workers (Fig. 2.11(b)). The net result is that the hydrolysis of *p*-nitrophenyl 3-*trans*-ferrocenylpropenoate is accelerated a millionfold by the cyclodextrin, a rate enhancement that is comparable with that produced by enzymes.

2.11 THE FUTURE

Improvements in biomimetic chemistry should accompany developments in elucidating the process of enzymic catalysis in terms of reaction mechanism.

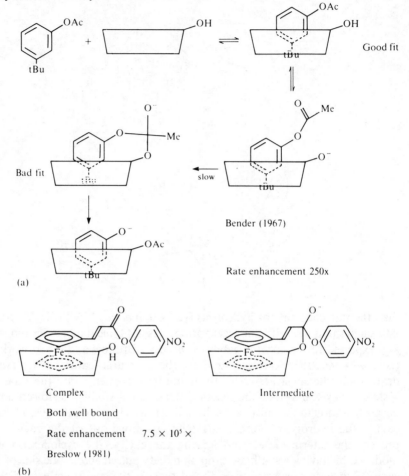

Bender (1967)

Rate enhancement 250x

(a)

Complex

Both well bound

Rate enhancement $7.5 \times 10^5 \times$

Breslow (1981)

(b)

Fig. 2.11 (a) The hydrolysis of *m,t*-butylphenyl acetate is facilitated by a cyclo-dextrin but steric problems prevent maximal efficiency. (b) A molecule designed to fit the cyclodextrin cavity in the catalytic transition state is subject to more rapid reaction.

Motion of the whole protein catalyst during a reaction, not just the active site, should also be taken into account. This is an extremely difficult problem to solve since there are few available methods that give specific information on macromolecular motion. We know that the protein is much more than simply a container for an active site. Furthermore, an active site cannot simply be a 'hat rack' for a collection of immobile functional groups. The most striking aspect of a protein chain is its relative flexibility. It is likely that the chain flexes to encourage reaction by following a series of local energy minima that arise in response to each step of the reaction sequence. This flexible interaction has been lacking in model enzymes that have so far been prepared. As a result, specificity and catalysis have been limited by the nature of the molecule which is modified to be the catalyst.

We have seen that some of the central problems involved in understanding enzymic catalysis can successfully be approached using the principles of physical organic chemistry. Although enzymes are not understood in mechanistic detail at the level of some simple organic reactions, we do know more about many enzymic reactions than we do about most organic reactions. Enzymes direct the substrate in a much more specific manner than is possible in an uncatalysed system. Therefore, the techniques of physical organic chemistry can be applied to enzyme reactions in a very productive way.

The interaction of substrates and enzymes remains one of the most fascinating mechanistic problems in modern organic chemistry. Yet our knowledge is still very limited. It is very doubtful that even the most accomplished organic chemist could examine the three-dimensional structure of an enzyme, whose function is not disclosed, and predict a reaction that the enzyme will catalyse efficiently. Designing a molecule that will have catalytic properties for useful reactions is a goal that is being approached so far with limited but significant success. We hope that the overview given in this chapter will help the reader who is new to the field appreciate the impact that enzyme chemistry has had upon other areas of chemistry.

REFERENCES

Abeles, R. H. and Maycock (1976) *Acc. Chem. Res.*, **9**, 313.
Albery, W. J. and Knowles, J. R. (1977) *Acc. Chem. Res.*, **10**, 105.
Aldersley, M. F., Kirby, A. J. and Lancaster, P. W. (1972) *J. Chem. Soc. Chem. Commun.*, 570.
Bell, R. P. (1973) *The Proton in Chemistry*, 2nd edn., Cornell University Press, Ithaca, New York.
Bender, M. L. (1957) *J. Am. Chem. Soc.*, **79**, 1258.
Bender, M. L. (1971) *Mechanisms of Homogeneous Catalysis from Protons to Proteins*, Wiley Interscience, New York.
Bender, M. L., Chow, Y. and Chloupek, F. (1958) *J. Am. Chem. Soc.*, **80**, 5380.
Benner, S. A., Rozzell, J. D., Jr. and Morton, T. H. (1981) *J. Am. Chem. Soc.*, **103**, 993.
Bernasconi, C. F., Hibdon, S. A. and McMurry, S. E. (1982) *J. Am. Chem. Soc.*, **104**, 3459.
Breslow, R. (1962) *Ann. N.Y. Acad. Sci.*, **98**, 445.
Breslow, R. and Wernick, D. (1976) *J. Am. Chem. Soc.*, **98**, 259.
Breslow, R., Czarniecki, M. F., Emert, J. and Hamaguchi, H. (1980) *J. Am. Chem. Soc.*, **102**, 762.
Bruice, T. C. (1970) *The Enzymes*, **2**, 217.
Bruice, T. C. and Benkovic, S. J. (1966) *Bioorganic Mechanisms*, W. A. Benjamin, New York.

Byers, L. D. (1978) *J. Theor. Biol.*, **74**, 501.
Cardinale, G. J. and Abeles, R. H. (1968) *Biochemistry*, **7**, 3970.
Cunningham, E. B. (1978) *Biochemistry*, McGraw-Hill, New York.
Curtin, D. Y. (1954) *Rec. Chem. Prog.*, **15**, 111.
Denbigh, K. (1964) *The Principles of Chemical Equilibrium*, Cambridge University Press, Cambridge.
Dunathan, H. C. (1966) *Proc. Natl. Acad. Sci. U.S.A.*, **55**, 712.
Dunathan, H. C. (1967) *Adv. Enzymol. Relat. Areas Mol. Biol.*, **35**, 79.
Eigen, M. (1964) *Angew. Chem. Int. Ed. Engl.*, **3**, 1.
Eliel, E. (1963) in *Stereochemistry of Carbon Compounds*, McGraw-Hill, New York, pp. 149–156.
Eliel, E. (1980) *J. Chem. Ed.*, **57**, 52.
Fersht, A. R. (1977) *Enzyme Structure and Mechanism*, W. H. Freeman and Co., San Francisco.
Gutowski, J. A. and Lienhard, G. E. (1976) *J. Biol. Chem.*, **251**, 2863.
Hall, A. and Knowles, J. R. (1975) *Biochemistry*, **14**, 4348.
Hammett, L. P. (1970) *Physical Organic Chemistry*, McGraw-Hill, New York.
Hammons, G., Westheimer, F. H., Nakaoka, K. and Kluger, R. (1975) *J. Am. Chem. Soc.*, **97**, 1568 and 4152.
Hanson, K. and Rose, I. A. (1975) *Acc. Chem. Res.*, **8**, 1.
Huber, C. P., Ozaki, Y., Pliura, D. H., Carey, P. R. and Storer, A. C. (1982) *Biochemistry*, **21**, 3109.
Ingold, C. K. (1953) *Structure and Mechanism in Organic Chemistry*, Cornell University Press, Ithaca, New York.
Jencks, W. P. (1969) *Catalysis in Chemistry and Enzymology*, McGraw-Hill, New York.
Jencks, W. P. (1975) *Adv. Enzymol. Relat. Areas Mol. Biol.*, **43**, 219.
Jencks, W. P. (1980) *Acc. Chem. Res.*, **13**, 161.
Kirby, A. J. (1980) *Adv. Phys. Org. Chem.*, **17**, 183.
Kirby, A. J. and Lancaster, P. W. (1972) *J. Chem. Soc. Perkin Trans.*, **2**, 1206.
Klotz, I. M. (1976) *J. Chem. Ed.*, **53**, 159.
Kluger, R. (1982) *Ann. N.Y. Acad. Sci.*, **378**, 63.
Kluger, R. and Adawadkar, P. D. (1976) *J. Am. Chem. Soc.*, **98**, 3741.
Kluger, R. and Chin, J. (1982) *J. Am. Chem. Soc.*, **104**, 2891.
Kluger, R. and Nakaoka, K. (1974) *Biochemistry*, **13**, 910.
Kluger, R. and Smyth, T. (1981) *J. Am. Chem. Soc.*, **103**, 214.
Kluger, R., Davis, P. P. and Adawadkar, P. D. (1979) *J. Am. Chem. Soc.*, **101**, 5995.
Krampitz, L. O. (1970) *Thiamin Diphosphate and its Catalytic Functions*, Marcel Dekker, New York.
Krampitz, L. O., Suzuki, I. and Greull, G. (1961) *Fed. Proc. Fed. Am. Soc. Exp. Biol.*, **20**, 971.
Lienhard, G. E. (1973) *Science*, **180**, 149.
MacInnes, I., Neshebel, D. C., Orszulik, S. T. and Suckling, C. J. (1982) *J. Chem. Soc., Chem. Commun.*, **121**, 1146.
MacNeil, P. A., Roberts, N. K. and Bosnich, B. (1981) *J. Am. Chem. Soc.*, **103**, 2273.
Metzler, D. E. (1977) *Biochemistry*, Academic Press, New York.
Murdoch, J. R. (1981) *J. Chem. Ed.*, **58**, 32.
O'Leary, M. H. and Baughn, R. L. (1972) *J. Am. Chem. Soc.*, **94**, 626.
Page, M. I. and Jencks, W. P. (1971) *Proc. Natl. Acad. Sci. U.S.A.*, **68**, 1678.
Ritchie, C. D. (1969) in *Solute–Solvent Interactions*, Vol. 1 (eds J. F. Coetzee and C. D. Ritchie), Marcel Dekker, New York.

Schowen, R. L. (1977) in *Isotope Effects on Enzyme-Catalyzed Reactions* (eds W. W. Cleland, M. H. O'Leary and D. B. Northrop), University Park Press, Baltimore, pp. 64–99.

Scrimgeour, K. G. (1977) *Chemistry and Control of Enzyme Reactions*, Academic Press, London.

Segel, I. H. (1975) *Enzyme Kinetics*, John Wiley and Sons, New York.

Tabushi, I. (1982) in *Frontiers of Chemistry* (ed. K. J. Laidler), Pergamon Press, Oxford, pp. 275–286.

Tagaki, W. and Westheimer, F. H. (1968) *Biochemistry*, **7**, 891.

Turano, A., Furey, W., Pletcher, J., Sax, M., Pike, D. and Kluger, R. (1982) *J. Am. Chem. Soc.*, **104**, 3089.

van Etten, R. L., Sebastian, J. F., Clowes, G. A. and Bender, M. L. (1967) *J. Am. Chem. Soc.*, **89**, 3242.

Walsh, C. (1979) *Enzymatic Reaction Mechanisms*, W. H. Freeman and Co., San Francisco.

Westheimer, F. H. (1956) *Adv. Enzymol. Relat. Subj. Biochem.*, **24**, 441.

Westheimer, F. H. (1969) *Methods Enzymol.*, **14**, 231.

Wittorf, J. H. and Gubler, C. J. (1970) *Eur. J. Biochem.*, **14**, 53.

Wolfenden, R. V. (1972) *Acc. Chem. Res.*, **5**, 10.

Wolfenden, R. V. (1974) *Mol. Cell. Biochem.*, **3**, 207.

Wolfenden, R. V. (1979) *FEBS Symp.*, **57**, 151.

Wood, H. G. and Barden, R. E. (1977) *Annu. Rev. Biochem.*, **46**, 385.

ACKNOWLEDGEMENT

This work has been supported by an operating grant from the Natural Sciences and Engineering Council of Canada.

3 | Chemical models of coenzyme catalyses

Seiji Shinkai

3.1 INTRODUCTION

An enzyme is born through repeated 'trial and error' in nature over an enormous length of time. Active sites of enzymes evolved to allow the enzymes to mediate biological reactions under ambient conditions and thus serve as excellent biological 'catalysts'. Nevertheless, enzymes have two inevitable limitations when viewed by organic chemists: firstly, enzymic catalyses are often too specific and do not allow the reactions that organisms do not require and secondly, the catalytic activities appear only under ambient physiological conditions. Although many useful synthetic reactions can be executed using enzymic catalysis (see Chapter 4), reaction conditions as well as enzyme specificity limit the direct applicability of enzymes. Therefore, one has to mimic the essence of the enzymic catalyses in more simplified systems in order to utilize the enzyme-like catalyses in a more versatile manner. This is the aim of an enzyme model study. We thus consider that the enzyme model study consists of two main targets: firstly, exploitation of more versatile, enzyme-like catalysts and secondly clarification of the reaction mechanisms in more simplified systems.

The specific concern of this chapter is with coenzymes which are prosthetic groups in many enzymes and bind to their apoenzymes via covalent bonds and/or secondary valence forces. Importantly, most coenzymes are capable of catalysing reactions, although weakly, even in the absence of apoenzymes, while apoenzymes take charge of enhancing the reactivity and controlling the stereoselectivity in the catalytic processes of coenzymes in the manner described in the previous chapter. Therefore, if one could control the reactivity and the stereoselectivity of coenzymes with 'artificial apoenzymes', these would become expedient and practical enzyme-modelled catalysts.

In this chapter, we review a number of efforts devoted to the application of coenzyme-catalysed systems to organic chemistry but will not refer to the detailed mechanisms of the coenzyme catalyses which have already been surveyed by other bio-organic chemists (Jencks, 1969; Bruice and Benkovic, 1966). The present review is the first comprehensive account of much coenzyme chemistry.

3.2 MODEL INVESTIGATIONS OF NICOTINAMIDE COENZYMES

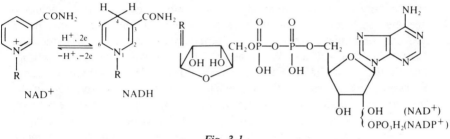

Fig. 3.1

The nicotinamide–adenine dinucleotide and its reduced form (NAD⁺ and NADH) serve as coenzymes in a large number of enzymic oxidation–reduction reactions (Fig. 3.1). NADH-dependent enzymes include the dehydrogenases, transhydrogenases, diaphorases, phosphorylases and oxidases. Although the enzymic mechanisms are not simple, we would like to consider, from a viewpoint of model studies, that these reactions simply take place by hydrogen exchange between the substrate and the pyridinium cation or 1,4-dihydropyridine in the nicotinamide moiety. The essential characteristics of NADH-dependent enzymes can be summarized as follows: (i) the reactions take place with the ternary (apoenzyme–coenzyme–substrate) complexes, (ii) the transfer of hydrogen is direct and apparently involves no exchange with solvent proton, and (iii) the reactions are stereospecific with respect to both the coenzyme and the substrate. These characteristic targets have been a great challenge for model investigations of NADH chemistry. Since the dihydronicotinamide moiety of NADH (or NAD⁺) is responsible for the redox reactions, 1,4-dihydronicotinamides (*3.1*) with simple *N*-substituents (e.g., R = *n* – Pr, benzyl, etc.) are used in model studies. Hantzsch esters (*3.2*) are also employed because of their preparative convenience.

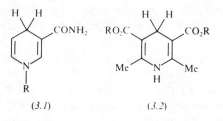

(3.1)   (3.2)

3.2.1 Metal ion catalysis

Alcohol dehydrogenase is one of the most representative NADH-dependent enzymes and has been a typical object of biomimetic NADH chemistry. In contrast to the ability of alcohol dehydrogenase to reduce aldehydes of widely varying structure (Jones and Beck, 1976), NADH model compounds reduce only the most activated carbonyl substrates such as hexachloro-acetone, trifluoroacetophenone, chloranil etc., a fact which has made drawing mechanistic conclusions from model reactions difficult. (Bruice and Benkovic, 1966; Kill and Widdowson, 1978; Sigman *et al.*, 1978; Dittmer and Fouty, 1964; Steffens and Chipman, 1971). Alcohol dehydrogenase is known to contain zinc at the active site which plays a decisive role in the mechanism of action. Pattison and Dunn (1976) suggested that the zinc ion serves as an electrophilic catalyst in the reduction process of the ternary complex (Fig. 3.2). The concept is very helpful to attain the biomimetic reduction of non-activated carbonyl substrates.

In 1971, Creighton and Sigman reported that 1,10-phenanthroline-2-carboxaldehyde *(3.3)* can be reduced by *(3.1)* in the presence of zinc ion. Probably, this was the first example of metal-assisted NADH model reduction of carbonyl groups. Subsequent studies have established that metal catalysis is quite versatile in the reduction of a variety of double bonds

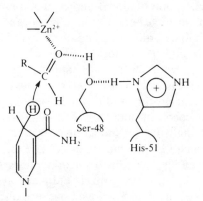

Fig. 3.2 Ternary complex of alcohol dehydrogenase.

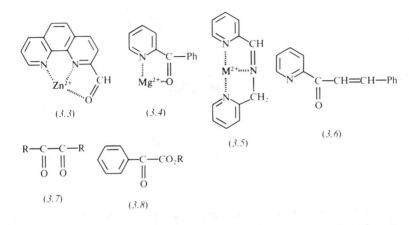

(3.3)   (3.4)   (3.5)   (3.6)

(3.7)   (3.8)

(C=O, C=N, C=C, etc.): for example, (3.4)–(3.8) were reduced by NADH model compounds in acetonitrile in the presence of metal ions (mainly $Mg(ClO_4)_2$) (Gase et al., 1976; Gase and Pandit, 1977; Shinkai et al., 1979a; Ohnishi et al., 1975a; Ohno et al., 1980).

Added metal ions sometimes divert the reaction pathway. Ohnishi et al. (1976) found in the reduction of (3.9) by (3.10) that both (3.11) and (3.12) are produced in the absence of Mg^{2+}, whereas (3.11) becomes a sole product in the presence of Mg^{2+} (Fig. 3.3). Since (3.12) is probably produced owing to dimerization of a radical intermediate, they considered that Mg^{2+} suppresses the radical nature of 1,4-dihydronicotinamide.

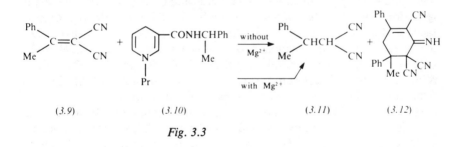

(3.9)   (3.10)   (3.11)   (3.12)

Fig. 3.3

3.2.2 Proton acid catalysis

It is known that the active sites of D-glyceraldehyde 3-phosphate dehydrogenase and lactate dehydrogenase (NADH-dependent enzymes) have an imidazole function of the histidine residue instead of zinc. X-ray-crystallographic studies (Adams et al., 1973; Moras et al., 1975) suggest that the protonated imidazole acts as a general acid during the reduction process (Fig. 3.4). The finding means that proton acids, in addition to Lewis acids

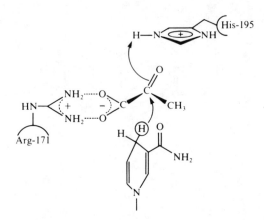

Fig. 3.4 Ternary complex of lactate dehydrogenase based on an X-ray crystallographic study (Adams *et al.*, 1973).

such as Zn^{2+} and Mg^{2+}, would also be useful as electrophilic catalysts in the NADH model reduction. It has been reported that the orthohydroxyl group plays a crucial role in the dihydronicotinamide reduction of the C=O, C=S and C=N double bonds attached to the aromatic ring in *(3.13)*–*(3.16)* (Pandit and Mas Cabré, 1971; Shinkai and Bruice, 1973a; Shinkai *et al.*, 1976a; Abeles *et al.*, 1957). Similarly, Shinkai and Kunitake (1977) noticed that the aldehyde group in *(3.17)* is readily reduced by *(3.1)*. These studies, which are model systems relevant to D-glyceraldehyde 3-phosphate dehydrogenase, demonstrate the importance of the hydrogen-bonding with the reaction centre of the substrate. Van Eikeren and Grier (1976) found that the reduction rate of trifluoroacetophenone by *(3.1:* R = *n* − Pr) is markedly facilitated in protic solvents rather than in aprotic solvents. It is also known that the addition of trace amounts of proton acids facilitates NADH model reductions (Shinkai and Kunitake, 1977; Wallenfels *et al.*, 1973). These results consistently support the idea that the transition state of NADH model reductions is favourably stabilized in protic or acidic media.

Here one should note the dilemma that conventional NADH model compounds rapidly decompose in acidic media to 1,4,5,6-tetrahydronicotinamide derivatives *(3.18)*. Thus, reduction in the presence of acid is always

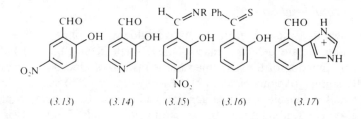

(3.13)   (3.14)   (3.15)   (3.16)   (3.17)

accompanied by the acid-catalysed decomposition of the NADH model compounds. To circumvent this undesired decomposition, we (Shinkai *et al.*, 1978a, 1979b) synthesized a new NADH model compound (*3.19*), in which the acid-sensitive 5,6-double bond of 1,4-dihydronicotinamide is protected by the aromatic ring. We found that (*3.19*) is 3.0×10^3 times more stable in acetic acid than (*3.1*: R = benzyl). With this acid-stable NADH model, we reduced inactivated carbonyl substrates (benzaldehyde, cyclohexanone, etc.), α-keto acids and α-imino acids (products are α-amino acids), in the presence of strong acids (HCl, benzenesulphonic acid, etc.) (Shinkai *et al.*, 1979c, 1979d, 1980a). We also synthesized an acid-stable NADH model (*3.20*) which reduces carbonyl substrates with the aid of an intramolecular acidic function (Shinkai *et al.*, 1979e).

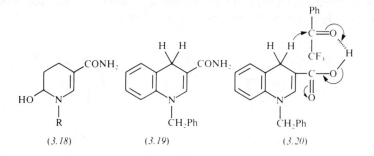

(3.18)   (3.19)   (3.20)

3.2.3 Catalysis through non-covalent interactions

Hadju and Sigman (1975) showed that dihydronicotinamides containing neighbouring carboxylate groups (*3.21*) can reduce *N*-methylacridinium ion in non-aqueous solution orders of magnitude more rapidly than can homologous derivatives lacking the free carboxylate group. For example, (*3.21*) exhibits a 100-fold rate acceleration relative to its methyl ester in acetonitrile; (*3.22*), which features the more conformationally inflexible carboxylate group, exhibits a still larger rate enhancement (10^3-fold) (Hadju and Sigman,

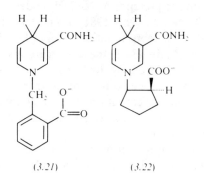

(3.21)   (3.22)

1977). The rate acceleration probably reflects the electrostatic stabilization of a positive charge which develops on a nicotinamide nitrogen in the reduction pathway.

Similar electrostatic stabilization may be anticipated for an NADH model compound (3.1: R = n – dodecyl) bound to anionic micelles. However, Shinkai et al. (1978b) reported that the rate constants for the reduction of N-methylacridinium ion and isoalloxazines are enhanced to smaller extents than in the intramolecular examples (3.21) and (3.22). On the other hand, the acid-catalysed decomposition of (3.1: R = n – dodecyl) is markedly accelerated by anionic micelles and suppressed by cationic micelles (Shinkai et al., 1975, 1976b; Bunton et al., 1978). The rate acceleration by anionic micelles is rationalized in terms of the electrostatic stabilization of a positively charged intermediate (3.23) by the micelle charge and the concentration of H_3O^+ on the micelle surface (Fig. 3.5).

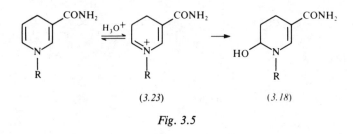

(3.23)   (3.18)

Fig. 3.5

3.2.4 NADH models as electron donors

NADH model compounds usually transfer hydride (or its equivalent) to substrates with the aid of metal ions or proton acids (MacInnes et al., 1982). On the other hand, they can act as stepwise one-electron donors when the substrates exhibit the characteristics of one-electron acceptors such as $Fe(CN)_6^{3-}$ and haemin (Kill and Widdowson, 1978). It is known that good leaving groups (e.g. NO_2^-, Cl^-, Br^-, $PhSO_2^-$, etc.) are easily eliminated from the sp^3-carbon in the presence of NADH model compounds (Kill and Widdowson, 1976; Ono et al., 1980, 1981; Inoue et al., 1974). For example, (3.24) undergoes reductive debromination via a radical intermediate (3.25) (Fig. 3.6). Similarly, in (3.26)–(3.28) the leaving groups (indicated by underlining) are reductively replaced by hydrogen. It is considered that the reactions take place by one-electron transfer from NADH model compounds via a nicotinamide radical cation. These reactions are closely akin to the S_{RN}-type reactions involving one-electron transfer followed by elimination of anionic groups.

An NADH model reduction that does not proceed thermally, sometimes takes place under photoirradiation. For example, dimethyl maleate and

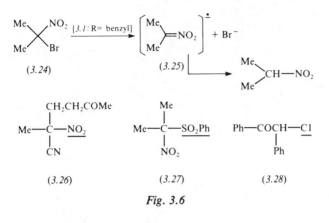

Fig. 3.6

fumarate are reduced to dimethyl succinate by (3.1: R = benzyl) or (3.2) under photoirradiation (> 350 nm), the yield being 60–70% (Ohnishi et al., 1975b). The reduction of C=C double bonds is further facilitated in the presence of Ru(bpy)$_3^{2+}$ (bpy = 2,2'-bipyridine) (Pac et al., 1981). The photo-activation of NADH model compounds is probably related to facile one-electron transfer at the excited state.

3.2.5   NADH models attached to host molecules

Biomimetic catalyses involving host–guest complexes of cyclodextrins have been reviewed in Chapter 2; host molecules bearing functional groups frequently mimic the essential functions of enzymic catalysis (Shinkai, 1982). They bind substrates to form the equivalent of a Michaelis complex which, as has been discussed, is considered to be the origin of high activity and selectivity of enzyme catalysis. For example, β-cyclodextrin bearing a NADH model function (3.29) has been synthesized (Kojima et al., 1981). It can reduce ninhydrin with a large rate enhancement (about 40-fold) relative to NADH and the reaction proceeds according to Michaelis–Menten-type saturation kinetics. Van Bergen and Kellog (1977) synthesized a crown ether NADH mimic (3.30) which is able to reduce a sulphonium ion $R^1COCH_2S^+R^2(CH_3)$ to R^2SCH_3, the increase in rate relative to (3.2) being 2700 times. On the other hand, the activity is efficiently quenched by NaClO$_4$. The marked rate enhancement is due to the interaction between the sulphonium ion and the crown ring, and the inhibition is due to the competitive binding of Na$^+$ to the crown ring. The phenomenon reminds us of competitive inhibition in enzyme chemistry. Similarly, the reactivity of NADH models immobilized in a cyclophane structure (3.31) was investigated by Murakami et al. (1981a, b),

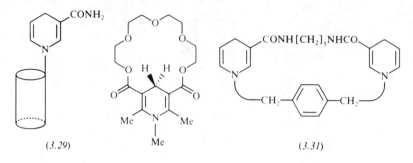

(3.29)

(3.30)

(3.31)

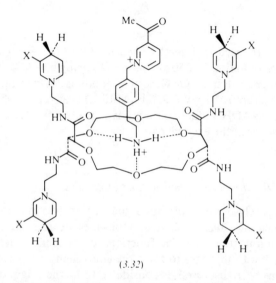

(3.32)

More specific design of NADH model catalysis has been reported by Behr and Lehn (1978). As illustrated in (3.32) the substrate is bound through the interaction between the crown ether of the catalyst and the ammonium group of the substrate, and hydrogen transfer occurs intramolecularly from dihydropyridine to the 3-acetylpyridinium ion. Examples of host–guest chemistry in synthesis will be found in the following chapter.

3.2.6 *Asymmetric reductions with NADH model compounds*

One of the most noteworthy characteristics of NADH-dependent enzymes is stereospecific hydrogen transfer, and, in considering asymmetric reductions, we come close for the first time in this chapter to a key interest of the synthetic chemist (see also Chapter 4). A mimic of the asymmetric reduction of a carbonyl substrate by a NADH model compound was first reported by

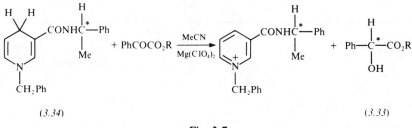

(3.34)   (3.33)

Fig. 3.7

Ohnishi *et al.* (1975c). They introduced a chiral α-phenethylamine to the 3-amido group of (*3.1*) and carried out the reduction of benzoylformic acid esters to mandelic acid esters (*3.33*) in the presence of Mg^{2+} ion (Fig. 3.7). The configuration of (*3.33*) was in accord with that of the 3-amido group, and the optical yields were 11−20%.

Considerable efforts to improve the optical yield have since been made. Representative examples are summarized in Table 3.1. The last three examples particularly warrant considerable attention because of the high enantioselectivity. In (*3.39*), the 4-position which releases a hydrogen is covered by an asymmetric crown ether, so that the orientation of the substrate is highly controlled in the transition state. In (*3.40*), the hydrogen used for reduction is linked directly to the asymmetric carbon. In (*3.41*), a Mg^{2+}-bridged cyclic structure is proposed as an active species. (*3.39*) and (*3.41*) are reusable by reducing the oxidized forms with sodium dithionite, whereas (*3.40*) is not because the reduction gives the racemic compound. These examples demonstrate that the *in vitro* enantioselectivity of the NADH model reduction is now as high as that of the enzymic systems.

3.2.7   Oxidation by NAD⁺ model compounds

In alcohol dehydrogenases, the interconversion of aldehydes (ketones) and alcohols occurs in conjugation with that of NADH and NAD^+ coenzymes. In contrast to a great number of investigations on the NADH model reduction of carbonyl substrates, there are few examples for the NAD^+ model oxidation of alcohol substrates. Wallenfels and Hanstein (1965) showed that 9-fluorenol is oxidized to fluorenone (8% yield) by (*3.42*). Shirra and Suckling (1977) have studied the oxidation of benzyl alkoxides to benzaldehydes with a variety of pyridinium salts (e.g. (*3.43*)), but the product analysis was not fully conducted. More recently, Ohnishi and Kitami (1978) carried out the oxidation of lithium alkoxides by (*3.44*), the yields of the 1,4-dihydropyridine being 3.5−28%. These unsatisfactory results suggest that the *N*-substituted pyridinium ions (conventional NAD^+ model compounds) have some defect as mimics of NAD^+-dependent oxidation, and model studies of oxidations followed an unexpected course.

Table 3.1 Optical yield for the asymmetric reduction of benzoylformic acid esters

Catalyst	Optical yield (%)	Reference
(3.34) Ph / CONHCHMe pyridine, N-R	11–20	Ohnishi *et al.* (1975c)
(3.35) R / CONHCHCONH$_2$ pyridine, N-CH$_2$Ph; R = Me, R = CH$_2$CHMe$_2$, R = CH$_2$Ph	47 26 5	Endo *et al.* (1977)
(3.36) RO$_2$C, CO$_2$R, Me, N-H, Me; R = (-)-menthyl	17–21	Nishiyama *et al.* (1976)
(3.37) CONHCHCH$_2$OH / CHOH / Ph; pyridine N-Pr	26.3	Makino *et al.* (1979)
(3.38) CONH$_2$; tBu—H; C$_6$H$_4$; OMe	27	van Ramesdonk *et al.* (1978)
(3.39) macrocycle, N-Me	86	de Vries and Kellog (1979)
(3.40) H Me Ph / CONH—C—Me / H; Me; N-Pr	94.7–97.6	Ohno *et al.* (1979)
(3.41) H$_2$NOC H ... CONH$_2$; CH$_2$—C$_6$H$_4$—CH$_2$	93–98	Seki *et al.* (1981)

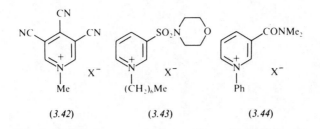

(3.42)   (3.43)   (3.44)

5-Deazaflavins (*3.45*) (and their reduced forms (*3.46*)) were at first synthesized as analogues of flavin coenzymes (O'Brien *et al.*, 1970; Edmondson *et al.*, 1972). Subsequent research established, however, that the redox behaviour of 5-deazaflavins is closely akin to that of NAD$^+$ rather than to that of flavin coenzymes (Hemmerich *et al.*, 1977). Surprisingly, it was later discovered that the basic skeleton of 5-deazaflavin is involved in fluorescent cofactor F$_{420}$ of *Methanobacterium* (Cheeseman *et al.*, 1972).

Yoneda *et al.* (1977) found that (*3.47*) is able to oxidize alcohols under alkaline conditions, the yields of aldehydes and ketones being 82–99%. The result indicates that the skeleton of 5-deazaflavin suitably mimics NAD$^+$ in the enzymes. Subsequently, they demonstrated that modified coenzyme models (*3.48*)–(*3.51*) also oxidize alcohols (Yoneda and Nakagawa, 1980; Yoneda *et al.*, 1978a, 1980a, b). Compound (*3.48*) shows strong oxidizing

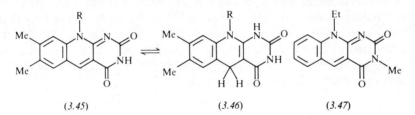

(3.45)   (3.46)   (3.47)

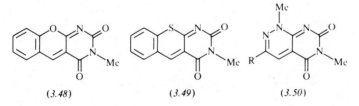

(3.48)   (3.49)   (3.50)

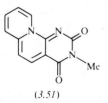

(3.51)

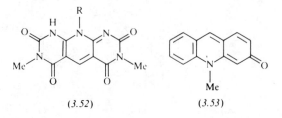

(3.52)   (3.53)

power owing to the electron-negative nature of the oxygen in the ring system, and oxidizes benzyl alcohol in the absence of base. The yields of the oxidation products per mole of heterocycle under aerobic conditions sometimes exceed 100% because the reduced forms are reoxidized by molecular oxygen. Yoneda *et al.* (1981) synthesized (3.52). As its reduced form is relatively sensitive to molecular oxygen, it acts as an excellent recycling oxidizing reagent. Shinkai *et al.* (1981a) designed another NAD$^+$ model (3.53) which is isoelectronic with 5-deazaflavin. We found that (3.53) efficiently oxidizes alcohols and the reduced (3.53) is reoxidized by molecular oxygen, the recycle number being 18–22.

Successful model NAD$^+$ oxidants all contain the equivalent of *N*-alkyl-pyridinium rings made electrophilic by substitution with several electron-withdrawing groups in a polycyclic system. Their efficacy compared with simple NAD$^+$ models can be ascribed to these structural features. Some of the model oxidants also have reducing properties. For example, it is known that reduced 5-deazaflavins are able to reduce aldehydes to alcohols in the presence of acids (Shinkai and Bruice, 1973b; Yoneda *et al.*, 1978b). It is interesting to note that the basic skeleton of an acid-stable NADH model (3.19) is involved in (3.46). This suggests that the reducing ability of (3.46) is also associated with its stability to acids.

3.3 FLAVIN CATALYSES

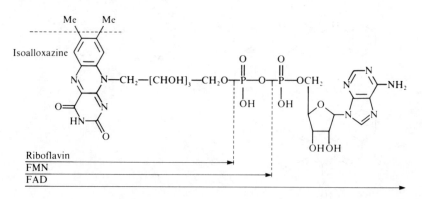

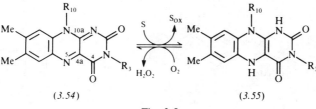

(3.54)   *(3.55)*

Fig. 3.8

Flavin coenzymes (FAD, FMN and riboflavin) oxidize a wide variety of substrates, such as amino acids, oxy acids, amines, etc., in biological systems,
and are themselves reoxidized by molecular oxygen (Hemmerich *et al.*, 1970;
Walsh, 1980; Bruice, 1980) (Fig. 3.8). From a chemical point of view, the
oxidation ability of flavins probably stems from the electron-withdrawing
triketone-like structure at the 4–4a–10a linkage which is comparable with
the reactive NAD$^+$ oxidant mimics just discussed. In simple α,β,γ-triketones
such as alloxane and ninhydrin, the oxidation ability is lost in aqueous
solution owing to facile hydration of the central carbonyl group. In contrast,
the triketone-like structure of flavins undergoes no hydration and retains its
strongly electron-deficient nature. Walsh *et al.* (1978) and Spencer *et al.*
(1977) investigated the importance of nitrogen atoms by systematically
replacing each nitrogen atom in the flavin with a carbon atom. They found
that the 1-deaza analogue *(3.56)* is as sensitive to oxygen as the native flavin
(3.55), but the 5-deaza analogues *(3.57)* and *(3.58)* are not (or scarcely) reoxidized by oxygen. This finding suggests that the sensitivity of reduced
flavin to oxygen is associated with the 5,10-enediamine structure.

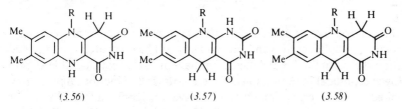

(3.56)   *(3.57)*   *(3.58)*

3.3.1   Flavins as oxidation catalysts

Flavin coenzymes serve as versatile oxidation catalysts in many biological
systems but, like the NAD$^+$ model systems, flavin molecules *in vitro* have a
rather small oxidation ability towards even relatively electron-rich substrates. It is known that flavins and isoalloxazines are able to oxidize
dihydropyridines such as *(3.1)* and *(3.2)*, aliphatic thiols and dithiols to
disulphides (Gibian and Winkelman, 1969; Gascoigne and Radda, 1967;
Gumbley and Main, 1976; Loechler and Hollocher, 1975), and carbanions
such as *(3.59)*–*(3.62)* (Main *et al.*, 1972; Rynd and Gibian, 1970; Brown and
Hamilton, 1970; Shinkai *et al.*, 1974). However, less-activated substrates

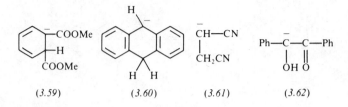

(3.59)   (3.60)   (3.61)   (3.62)

such as thiophenol and nitroalkane carbanions are stable to simple flavin models.

There are now recognized two main methods for 'activating' flavin molecules as oxidizing catalysts: (i) shifting the redox potential of flavins to more positive values, and (ii) converting substrates into more 'specific' ones with the aid of second cofactors or environmental effects. Method (i) is exemplified by $(3.63)-(3.65)$ (Yokoe and Bruice, 1975; Bruice *et al.*, 1977; Knappe, 1979; Shinkai *et al.*, 1979f), in which the electron-withdrawing substituents enhance the electron-deficiency of the isoalloxazine ring. For example, (3.63) is able to oxidize thiophenol and 2-nitropropane carbanion under ambient conditions (Yokoe and Bruice, 1975).

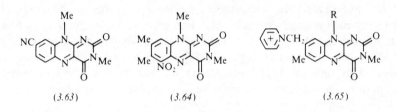

(3.63)   (3.64)   (3.65)

Flavin coenzymes serve as electron carriers to and from iron–sulphur proteins, molybdenum (xanthine oxidase) and haem proteins, and the interactions between flavins and metal ions have attracted considerable attention. However, model investigations of metal–flavin interactions have been very limited. The reason is simple: flavins have no significant affinity for metal ions except for a few such as Ag(I), Cu(I), Mo(V), Ru(II) and Fe(II) (Fritchie, 1972; Bamberg and Hemmerich, 1961; Clarke *et al.*, 1979; Selbin *et al.*, 1974). The stability of these metal–flavin complexes is probably due to a metal→flavin charge transfer. To overcome this deficiency, Shinkai *et al.* (1982a, 1983) synthesized a metal-co-ordinative flavin (3.66) which combines the structure of a flavin with the well-known chelate 1,10-phenanthroline and binds various kinds of metal ions. The metal–(3.66) complex acts as a strong oxidizing agent owing to the strong electron-withdrawing nature of metal ions; in particular, the Zr^{4+} complex is capable of oxidizing alcohols at room temperature. This reaction can be classified as an example of method (i) of activating a flavin.

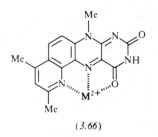

(3.66)

Method (ii) 'activates' substrates in the same way as apoenzymes do in the apoenzyme–coenzyme–substrate ternary complexes. The most expeditious method is to utilize the hydrophobic environments of micelles and polymers, where anions are significantly activated (Kunitake and Shinkai, 1980). In contrast to the inability of conventional flavins to oxidize thiophenol and nitroalkane carbanions in solution, hydrophobic (3.67) and (3.68) bound to a cationic micelle can oxidize these substrates under mild conditions (Shinkai et al., 1980b). Probably, both micellar activation and concentration of the substrate facilitates the oxidation reaction in the micelle, suggesting the importance of the hydrophobic environments. Similarly, flavins immobilized in cationic polymers act as oxidation catalysts for thiophenol and nitroethane carbanion, and the oxidation of NADH proceeds according to Michaelis–Menten saturation kinetics owing to the electrostatic interaction between cationic polymers and anionic NADH (Shinkai et al., 1978c, d, e; Spetnagel and Klotz, 1978).

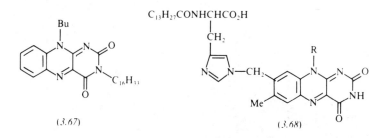

(3.67)   (3.68)

It was proposed on the basis of the oxidation of β-chloroalanine by D-amino acid oxidase (Walsh et al., 1971) that some flavoenzymes do not oxidize substrates directly but the conjugate base generated from the substrates (Kosman, 1977). The concept is very helpful for mimicking flavin oxidation in a non-enzymic system. For example, neither aldehydes nor α-keto acids are oxidized by flavins, but in the presence of cyanide ion these substrates are converted to cyanohydrin carbanions (3.69) and undergo rapid flavin oxidation to give the corresponding carboxylic acids (Shinkai et al., 1980c)

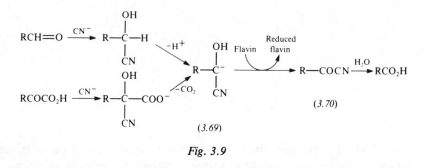

Fig. 3.9

(Fig. 3.9). Similarly, thiazolium ions, which exhibit a catalytic activity closely resembling that of cyanide ion, form 'active aldehyde' intermediates (3.71) from aldehydes and α-keto acids, and (3.71) is rapidly oxidized by flavins (Shinkai et al., 1980d, e; Yano et al., 1980) (Fig. 3.10). Interestingly, this reaction model well mimics the catalytic action of pyruvate oxidase which converts pyruvic acid to acetic acid with the aid of FAD and thiamin pyrophosphate. The reactions starting from α-keto acids are examples of the decarboxylative oxidation. Since (3.70) and (3.72) have active acyl groups, the reaction in alcohols gives the corresponding esters (Shinkai et al., 1980e; Yano et al., 1980). Shinkai et al. (1982b) synthesized a biscoenzyme (3.73) which has both flavin and thiazolium ion within a molecule. (3.73) efficiently catalyses the reaction sequence as intramolecular processes.

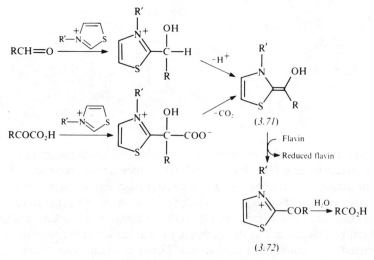

Fig. 3.10

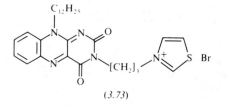

(3.73)

3.3.2   Flavin–oxygen complexes as oxygenase models

Reduced flavins (3.56) are rapidly reoxidized by molecular oxygen to yield hydrogen peroxide, but they can reduce pyruvate esters, nitrobenzene and quinones under anaerobic conditions (Williams and Bruice, 1976; Gibian and Baumstark, 1971; Gibian and Rynd, 1969). On the other hand, (3.74), in which N-5 of the flavin skeleton is blocked by a methyl group, produces the radical species (3.75) by one-electron oxidation, and 4a-hydroperoxyflavin (3.76) by O_2 addition (Kemal and Bruice, 1976a; Kemal et al., 1977) (Fig. 3.11). It is believed that the structure of (3.76) is similar to enzymically generated hydroperoxyflavin. 4a-Hydroperoxyflavin (3.76) transfers oxygen or dioxygen to sulphides (3.77) and (3.78), amines (3.79), phenolate ions (3.80), indoles (3.81), etc. (Kemal et al., 1977; Miller, 1982; Ball and Bruice, 1980; Kemal and Bruice, 1979; Muto and Bruice, 1980) (Fig. 3.12). Thus, these reactions become good biomimetic examples of mono- and di-oxygenases.

Meanwhile, Rastetter and co-workers (1979; Frost and Rastetter, 1981) have demonstrated that a flavin N(5)-oxide (3.82) also transfers mono-oxygen to phenols, amines, hydroxylamines, etc. They consider that in

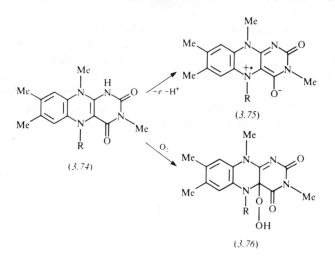

Fig. 3.11

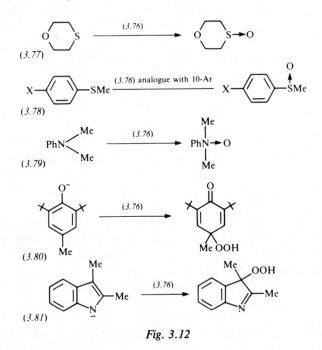

Fig. 3.12

mono-oxygenases, 4a-hydroperoxyflavin may be converted to flavin $N(5)$-oxide prior to oxygen transfer, and a flavin $N(5)$-nitroxyl radical is a viable candidate for the ultimate hydroxylation agent in the enzyme-catalysed reaction. A mechanistic understanding of flavin-mediated oxygenation is still incomplete.

A more simplified and synthetically important oxygenase model was offered by Heggs and Ganem (1979). As described previously, oxidized flavins have a triketone-like structure. In other words, the 4a-hydroperoxide in (3.76) is flanked by two electron-withdrawing groups and lies adjacent to a weakly basic, electronegative group (5-NH). They pointed out that (3.83), which is readily derived from hydrated hexafluoroacetone and hydrogen peroxide, completely satisfies these criteria; (3.83) acts as a low-cost, catalytic epoxidation reagent.

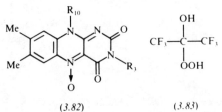

(3.82)   (3.83)

3.3.3   Model studies of bacterial luciferase

Another fascinating aspect of flavin chemistry is a chemiluminescence phenomenon imitating the bioluminescence of bacterial luciferase. It has been shown that the key step of bioluminescence catalysed by bacterial luciferase is the reaction of luciferase-bound 4a-hydroperoxyflavin mononucleotide (*3.84*) with a long-chain aldehyde (Hastings *et al.*, 1973) (Fig. 3.13).

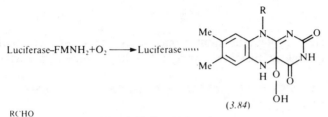

(*3.84*)

Fig. 3.13

Kemal and Bruice (**1976b**) found that (*3.76*) undergoes a chemiluminescent reaction in the presence of an aldehyde which was visible to the dark-adapted eye. Similar light emission was observed when a flavinium salt (*3.85*) was mixed with alkyl hydroperoxides, the quantum yields being 3.0×10^{-4}–3.3×10^{-4} (Kemal and Bruice, 1977) (Fig. 3.14). They proposed a structure (*3.86*) common to chemiluminescent species.

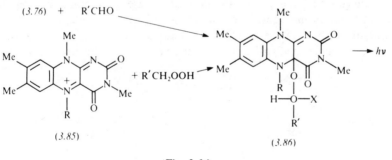

(*3.85*)

(*3.86*)

Fig. 3.14

McCapra and Lesson (1976) also found chemiluminescence for the reaction of (*3.87*) and peroxides (*3.88*), the quantum yields being 5×10^{-6}–2.1×10^{-5} (Fig. 3.15). They consider that, in this case, the addition of the peroxides occurred at position 10a.

These results, together with those of oxygen-transfer reactions, suggest that it is essential to substitute a labile $N(5)$-hydrogen with alkyl groups to

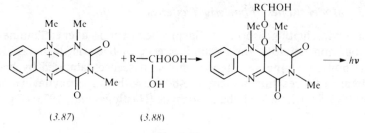

(3.87)   (3.88)

Fig. 3.15

imitate the flavoprotein functions. Conceivably, the $N(5)$-hydrogen of flavin coenzymes is tightly 'fixed' by hydrogen-bonding in the enzyme active sites.

3.4   CATALYSES RELATING TO VITAMIN B₁ AND ANALOGUES

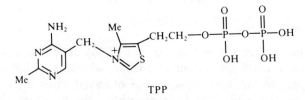

TPP

Thiamin pyrophosphate (TPP) and its analogues (3.89) (Vitamin B_1 family) serve as catalysts for condensation, decarboxylation and reduction reactions. In 1958, Breslow showed that the catalytic role of TPP can be ascribed to the 2-H deprotonated conjugate bases of a thiazolium ion (3.90), which behaves analogously to cyanide ion in benzoin condensation. It is now recognized that, in the reaction sequence of thiazolium ion catalysis, the active aldehyde (3.91) always plays the role of a key intermediate (Fig. 3.16).

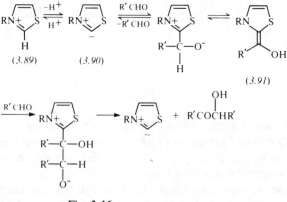

Fig. 3.16

3.4.1   Thiazolium ions as condensation catalysts

Breslow (1958) demonstrated that a variety of thiazolium ions are active in alcohol solvents in the presence of base as acyloin condensation catalysts unless position 2 is blocked by alkyl groups. The potential of the condensation reaction stems from the strong nucleophilicity of the active aldehyde (*3.91*). Cookson and Lane (1976) utilized (*3.89*: R = benzyl) to synthesize cyclic 2-hydroxy-2-enones (*3.92*) from dialdehydes; this is an example of an intramolecular acyloin condensation. Compound (*3.91*) is able to attack not only the aldehyde group but also nitrosobenzene (*3.93*), disulphides (*3.94*), ethyl acrylate and acrylonitrile (*3.95*: X = COOEt, CN) and vinyl alkyl ketones (*3.96*) (Corbett and Chipko, 1980; Rastetter and Adams, 1981; Stetter *et al.*, 1980; Stetter and Kuhlmann, 1976) (Fig. 3.17).

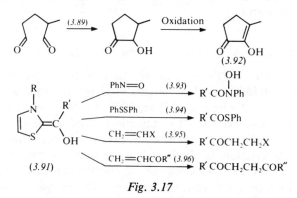

Fig. 3.17

In order to use the thiazolium ions as synthetic reagents, Castells *et al.* (1978) prepared a thiazolium ion immobilized in insoluble polymer supports. However, in contrast to the activity of thiazolium ions in alcohol solvents, they are totally inactive in aqueous solution. This is probably due to OH^--catalysed decomposition of the thiazolium ring which takes place in precedence over 2-H deprotonation. Tagaki and Hara (1973) found that surfactant thiazolium ions such as *N*-dodecylthiazolium bromide are very active in the aqueous acyloin condensation. Similarly, thiazolium ions covalently linked to cationic polymers are active in aqueous solutions (Shinkai *et al.*, 1982c). They consider that cationic environments facilitate the deprotonation of 2-hydrogen, which leads to the enhancement in the catalytic activity.

Another interesting mimic of thiazolium ion catalysis is an asymmetric synthesis. Since acetoin and benzoin, the products of the thiazolium-catalysed condensation, have a chiral carbon, one may expect that the condensation could proceed in a stereoselective manner. Sheehan and Hara (1974) reported that thiazolium ions with chiral *N*-substituents (*3.97*) give

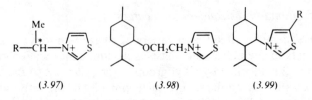

(3.97)   (3.98)   (3.99)

optically active benzoins in methanol. In particular, the optical yield in the presence of (3.97: R = p − O₂N-Ph) was significantly high (51.5%). Tagaki et al. (1980) also examined the asymmetric benzoin condensation under micellar conditions. They found on the chemical yields of benzoin that the catalytic activity is high for (3.98) and low for (3.99), but (3.99) gave much higher optical yield (35.5%) than did (3.98) (1.8–3.5%). These results suggest that the chiral centre of the N-substituent should be bulky and close to the reaction centre.

3.4.2   Thiazolium ions as decarboxylation catalysts

Another important TPP-mediated reaction is the decarboxylation of pyruvate to acetaldehyde which is catalysed by pyruvate decarboxylase (Crosby et al., 1970; Kluger and Pike, 1979; Kluger et al., 1981; see also Chapter 2). We have mentioned that thiamin catalysis in the acyloin condensation is closely akin to the cyanide ion catalysis in the benzoin condensation; decarboxylation is similar. The mechanisms proposed for the thiamin-catalysed decarboxylation also involve the key intermediate (3.91). One can write a reaction sequence equivalent to the cyanide ion catalysis which involves a cyanohydrin carbanion (3.69) as the key intermediate analogue (Fig. 3.18). Indeed many years ago, Franzen and Fikentscher (1958) found that cyanide ion acts as a good catalyst for the decarboxylation of α-keto acids.

Here, one should pay attention to one important characteristic of decarboxylation reactions, namely that they are very sensitive to solvent effects. In general, the reactions are fast in dipolar aprotic solvents and are markedly

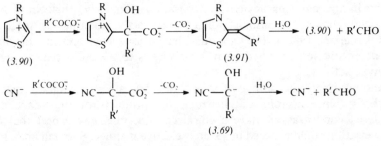

Fig. 3.18

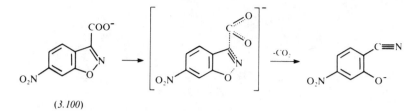

(3.100)

Fig. 3.19

suppressed in protic solvents owing to solvation of the carboxylate anion through hydrogen-bonding (Kunitake and Shinkai, 1980) (Fig. 3.19). For example, the rate constant for the decarboxylation of (3.100) in dimethyl formamide is greater by 5.1×10^6-fold and 1.5×10^5-fold than those in water and methanol respectively (Kemp and Paul, 1970). If some part of the enzyme catalysis is reproduced by the solvent effect, this result implies that the enzymic decarboxylation would be markedly facilitated in a water-free, hydrophobic active site of an enzyme, as was discussed in Chapter 2.

As expected, thiazolium ion-mediated decarboxylation of α-keto acids does not occur, or hardly, in aqueous solution (Crosby et al., 1970; Shinkai et al., 1980e). On the other hand, the cationic-micelle-bound thiazolium ion (3.89: $R = n - C_{16}H_{33}$) catalysed the decarboxylation of α-keto acids under ambient conditions. This result suggests that the hydrophobic microenvironment of the micelle is favourable to the decarboxylation and is in line with the fact that the cationic micelle accelerates the decarboxylation of (3.100) by a factor of 90 (Bunton et al., 1973). However, the solvent effect on the thiazolium ion-mediated decarboxylation is not fully understood.

3.4.3   Thiazolium ions as reducing agents

The key intermediate (3.91) which usually acts as a strong nucleophile becomes a reducing agent for one-electron-accepting substrates. Christen and Gasser (1980) and Cogoli-Greuter et al. (1979) demonstrated the irreversible inactivation of pyruvate decarboxylase and transketolase by hexacyanoferrate (III) or H_2O_2. These TPP-dependent enzymes involve the TPP

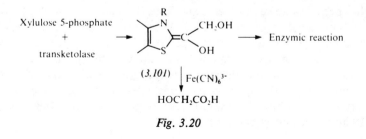

Fig. 3.20

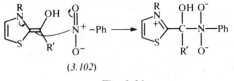

(3.102)

Fig. 3.21

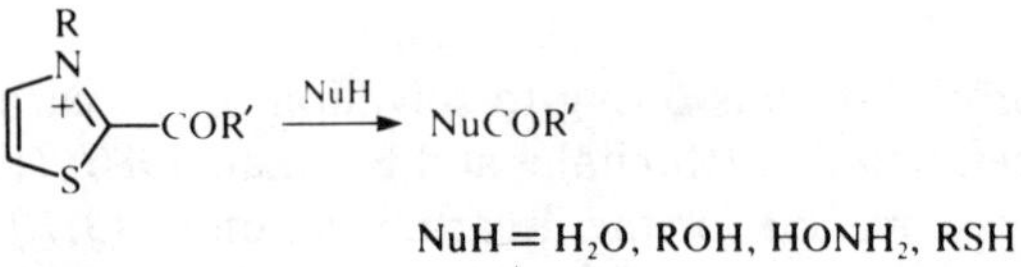

R
N
$\downarrow$
$\overset{+}{N}$—COR' $\xrightarrow{\text{NuH}}$ NuCOR'
S

(3.103)

NuH = H_2O, ROH, $HONH_2$, RSH

Fig. 3.22

analogues of (3.91) in their reaction sequences, which are oxidatively trapped by a one-electron oxidant. For example, transketolase produces the dihydroxyethyl intermediate (3.101), which is rapidly oxidized by hexacyanoferrate (III) to glycollic acid (Fig. 3.20). The result clearly suggests that the key intermediate (3.91) can act not only as a nucleophile but also as a reducing agent.

As has been described in the section on flavin catalysis, analogues of (3.91) rapidly reduce flavins (Shinkai et al., 1980d, e, 1982b; Yano et al., 1980). Inoue and Higashiura (1980) found that acridine and 10-methyl-acridinium chloride are reduced to 9,9′-biacridan and 10,10′-dimethyl-9,9′-biacridan respectively by (3.91: R = benzyl) plus base in methanol. These examples suggest that (3.91) has the character of a one-electron donor. Castells et al. (1977) employed p-nitrobenzaldehyde as a substrate for the thiazolium ion-mediated condensation of aldehydes. Unexpectedly, they could not detect 4,4′-dinitrobenzoin but the self-oxidoreduction products such as nitro-, azo- and azoxy-benzoic acids and esters. This finding implies that the nitro group serves as an oxidant of (3.91). They proposed a nucleophilic attack of (3.91) on the nitro group (3.102), but the possibility of an electron-transfer mechanism cannot be ruled out (Fig. 3.21).

It has been established by Daigo and Reed (1962) and White and Ingraham (1962) that 2-acylthiazolium ions (3.103) are easily cleaved by nucleophiles: for example, (3.103) gives carboxylic acids by water, esters by alcohols, hydroxamic acid by hydroxylamine, and thioesters by mercaptans. In other words, (3.103) is a versatile acylation reagent (Fig. 3.22). Since (3.103) can be obtained from thiazolium salts + aldehydes in one step, one can easily convert aldehydes into, for example, esters (Shinkai et al., 1980e; Yano et al., 1980; Inoue and Higashiura, 1980).

3.5 PYRIDOXAL CATALYSES

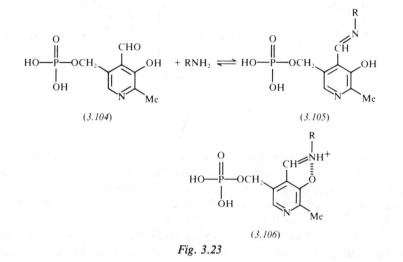

Fig. 3.23

It is well known that pyridoxal (vitamin B_6) catalyses proceed via a Schiff base and its tautomeric form (*3.106*) (Metzler *et al.*, 1954) (Fig. 3.23). Pyridoxal-dependent enzymes catalyse a broad range of reactions of amino acids such as decarboxylation, elimination, transamination, etc. Although the mechanistic aspects of these reactions have been studied, the impact on organic chemistry is rather limited. For instance Llor and Cortijo (1977) showed that the tautomerism between (*3.105*) and (*3.106*) is useful as a measure of the solvent polarity. These two tautomeric forms have different absorption maxima (415 and 335 nm respectively), and the relative strength changes depending on the solvent polarity. They confirmed that ΔG values for the equilibrium are linearly correlated to Kosower's Z values. This model becomes an especially useful strategy for measuring the polarity of the active sites of pyridoxal-dependent enzymes.

Buckley and Rapoport (1982) have demonstrated an interesting synthetic application of a pyridoxal mimic: in the presence of base 4-formyl-1-methyl-pyridinium ion acts as a convenient reagent for the chemical modification of primary amines to aldehydes and ketones, the yields being relatively good (77–94%). This process is not only interesting from the viewpoint of bio-mimetic chemistry but it is also important as a simple alternative to such transformation procedures.

Breslow *et al.* (1980) attached the pyridoxamine unit to β-cyclodextrin and examined the possibility of substrate specificity in transamination and chiral induction. While pyridoxamine itself exhibits similar reactivity for three α-keto acids (*3.108*: R = Me, benzyl, CH_2-indole), (*3.107*) catalyses the

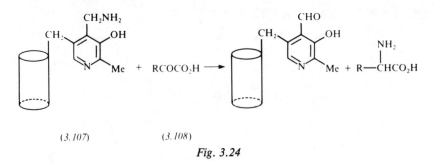

(3.107)   (3.108)

Fig. 3.24

transamination with indolepyruvic acid (*3.108*: R = CH$_2$-indole) about 200
times more efficiently than in the absence of the cyclodextrin (Fig. 3.24). The
rate acceleration due to the neighbouring cyclodextrin was not observed for
pyruvic acid (*3.108*: R = Me). Clearly, the difference is due to the binding of
the indole moiety into the cavity of β-cyclodextrin. Because the cavity of
β-cyclodextrin is chiral, one might expect some chiral induction in the product α-amino acids. Breslow's group reported that dinitrophenyltryptophan
and dinitrophenylalanine have 12% and 52% excess of the L-isomer
respectively. Hence, this 'artificial pyridoxal enzyme' shows significant
optical induction.

Kuzuhara *et al.* (1978) synthesized a pyridoxal analogue (*3.109*) involved
in an asymmetric cyclophane structure. It was found that the phane derivatives with the ring size equal to or less than fourteen members can be optically resolved into enantiomers (i.e., *n* = 4,5,6 in (*3.109*)) (Iwata *et al.*,
1976). The stereospecificity in the pyridoxal-like catalysis was tested in
racemization of L- and D-glutamic acids with (*3.109*: *n* = 5, (–)-form) in the
presence of cupric ions (Kuzuhara *et al.*, 1977) but the L-isomer was found to
racemize only 1.3 times faster than the D-isomer.

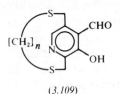

(3.109)

Micellar effects on pyridoxal catalysis were examined by using αβ-
elimination of *S*-phenylcysteine from the Schiff base (*3.110*) (Murakami and
Kondo, 1975) (Fig. 3.25). While anionic and non-ionic micelles showed little
or somewhat retarding catalytic effects, the cationic CTAB micelle pro-
moted the elimination by a factor of 7.1. Probably, the catalytic efficiency of
the CTAB micelle stems from the enhanced concentration of OH$^-$ which
abstracts α-hydrogen to induce the αβ-elimination. Similarly, Nakano *et al.*

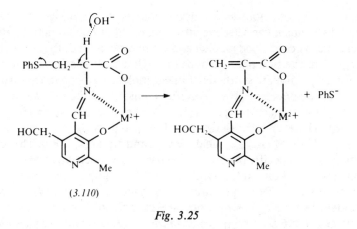

(3.110)

Fig. 3.25

(1981) have found that the pyridoxal-promoted $\alpha\beta$-elimination of serine is facilitated by metal ion plus quaternized poly (4-vinylpyridine). The contribution of the electrostatic force between the polymer and the metal complex of the Schiff base was suggested to account for the rate stimulation. Such elimination reactions have been used as the basis for the design of enzyme inhibitors (Chapter 5).

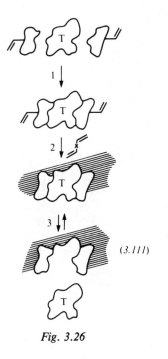

(3.111)

Fig. 3.26

One of the recent noteworthy developments in the design of enzyme-stimulated systems is the selective introduction of organic functionality in a stereochemically controlled manner and the maintenance of the stereochemical integrity of the functional groups through a synthesis (see also Chapter 4). By utilizing template synthesis of macromolecules through the insolubilization process, one may give a specific 'memory' of the original template ('T') to the 'ghost' macromolecules (*3.111*) (Shea *et al.*, 1980; Shinkai, 1982) (Fig. 3.26). These investigations are still in progress (Nishide *et al.*, 1976; Wulff *et al.*, 1977; Shea *et al.*, 1980: Damen and Neckers, 1980). The scheme reminds us of the specific binding of coenzymes to apoenzymes or substrates to enzymes. Belokon and coworkers (1980) applied this concept to a pyridoxal system. They prepared a polymeric gel (*3.114*) in which salicylaldehyde and lysine moieties are capable of forming an internal aldimine at pH > 6.0 by co-polymerization of (*3.112*) with acrylamide and (*3.113*) with the subsequent removal of the copper ions by EDTA (Fig. 3.27). They found that the equilibrium formation constants of the 'internal' aldimine are *ca.* 30 times (pH 7.1) and *ca.* 100 times (pH 9.2) higher than that of the model reaction. The result suggests that the 'memory' to form the aldimine is retained in the insolubilized polymer.

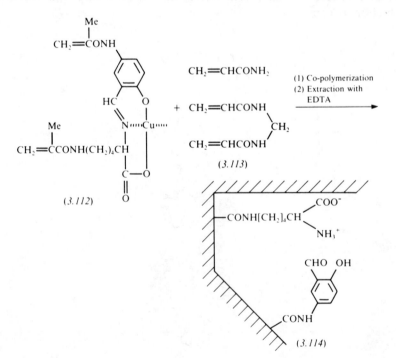

Fig. 3.27

Another important pyridoxal-mediated reaction is decarboxylation, which is, as mentioned previously, highly dependent on reaction media. One must therefore take the solvation of the carboxylate group into consideration in order to mimic the decarboxylase catalysis. It has been reported that pyridoxal-mediated decarboxylation of amino acids occurs under rather drastic conditions (100°C, 4h) (Kalyankar and Snell, 1962), but no example of a model reaction under ambient conditions exists.

3.6 CATALYSES OF THIOL COENZYMES

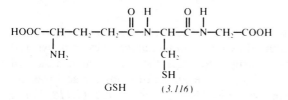

Lipoic acid (*3.115*)

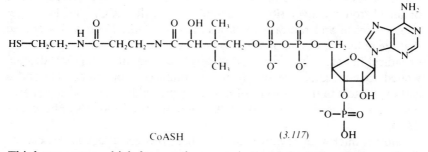

GSH   (*3.116*)

CoASH   (*3.117*)

Thiol coenzymes which frequently appear in the biological systems are lipoic acid (*3.115*), glutathione (GSH: *3.116*) and coenzyme A (CoASH: *3.117*). Although these coenzymes catalyse a broad range of reactions, the catalyses may be reduced to the following classification from the viewpoint of organic chemistry: thiol coenzymes act as (i) strong nucleophiles as well as good-leaving groups, (ii) mediators of oxidoreduction reactions involving group transfers, and (iii) electron-withdrawing groups to facilitate the ElCB-type deprotonation. In this article, we do not refer to (i) (see Kunitake and Shinkai, 1980).

Lipoic acid is widely distributed among micro-organisms and is frequently coupled to TPP: that is, it receives an acyl unit from (*3.91*) and transfers it to CoASH (Fig. 3.28). A model reaction has been reported by Rastetter and Adams (1981) (see Section 4.1). Takagi *et al.* (1976) utilized the redox couple

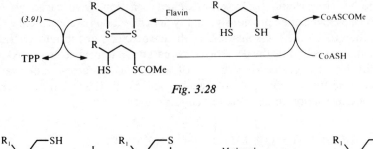

Fig. 3.28

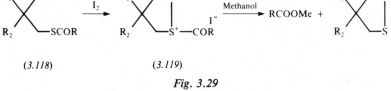

(3.118)   (3.119)

Fig. 3.29

between dihydrolipoic acid and lipoic acid to the activation of acyl groups. Structure (3.118) when oxidized by iodine in methanol gives an active acyl intermediate (3.119), which easily undergoes the nucleophilic attack of solvent methanol (Fig. 3.29). This is an example of oxidative acyl transfer and a facile method for ester synthesis. They later found that (3.118) can be obtained from aldehyde (RCHO) and disulphide ($R_1R_2CCH_2SSCH_2$) under photochemical and radical conditions (Takagi et al., 1980) and proposed that the reductive fission of the S—S linkage occurs by means of the photo- or radical-initiated chain reaction. Meanwhile, Nambu et al. (1981) demonstrated that an analogue of (3.118) is produced from lipoamide and a carboxylic acid in the presence of tri-n-butylphosphine. They also prepared lipoic acid immobilized in cross-linked polystyrene beads (Nambu et al., 1980).

Another interesting feature is the fact that hydrogen attached to α-carbon of thioesters is easily removed as a proton

$(coenzyme-SCOCH_2R \xrightarrow{-H^+} coenzyme-SC\bar{O}HR$: Yaggi and Douglas, 1977).

This is a pre-equilibrium step for a subsequent condensation. The synthetic utility of thiol ester enolate anions has been explored by Wilson and Hess (1980). For example, chain elongation to (3.120) and cyclization to (3.121) are possible by the reaction of thioesters and base via the carbanionic intermediates (Fig. 3.30).

It has been established by Oae et al. (1961) that the sulphur atom is able to stabilize the neighbouring carbanion, and the fact became an important clue to clarifying the reaction mechanism of glyoxalase I. The mechanism by which the glyoxalase I (GSH-dependent enzyme) catalyses the conversion of glyoxals to corresponding α-hydroxy acid esters has been a controversial

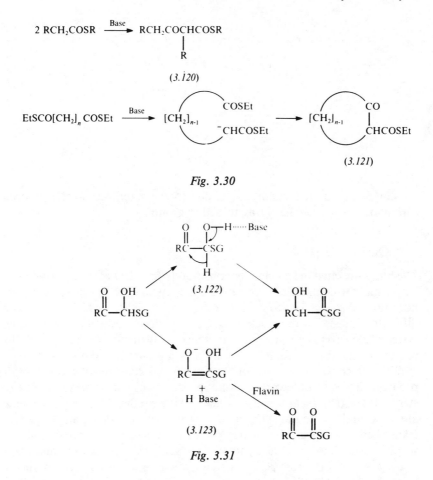

Fig. 3.30

Fig. 3.31

problem. Racker (1951) and Franzen (1955) proposed the intramolecular 1,2-hydride shift (*3.122*) like the Cannizzaro reaction, but the possibility of the enediol mechanism (*3.123*) has not been excluded (Fig. 3.31). Recently, the enediol mechanism was supported by several independent experiments: (i) a solvent proton is incorporated into the product α-hydroxy acid ester (Hall *et al.*, 1978), (ii) the enediol intermediate (*3.123*) is rapidly trapped by a flavin which does not serve as an efficient oxidant for intermediates including the 1,2-hydride shift mechanism (*3.122*) (Shinkai *et al.*, 1981b) (Fig. 3.31), (iii) fluoromethylglyoxal (*3.124*) is converted to pyruvic acid via the β-elimination path (Kozarich *et al.*, 1981) (Fig. 3.32), and (iv) the substance with the enediol structure as a transition-state analogue inhibitor (Douglas and Nadvi, 1979) (see Chapter 5). In conclusion, the acidity of hemithiol acetals as carbon acids is primarily due to the electron-withdrawing nature of

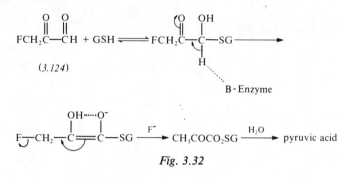

Fig. 3.32

the acyl group and, in addition, is due to the stabilization of produced carbanions by the neighbouring thioether group.

3.7 CONCLUSION

The literature cited in this chapter consistently reveals that the reactivities of coenzymes are controlled by solvent effects, metal ions, micelles, polymers, host molecules, etc. The control of the stereoselectivities, which seems more difficult than that of the reactivities, has also been attained in some model systems. However, one may notice that it is still difficult to enhance both the reactivity and the stereoselectivity of model systems at the same time, a facility that enzymes possess. Further efforts should be devoted toward this problem. It is also noteworthy that applications of coenzyme chemistry to synthesis are growing apace. Nevertheless it is clear that the successes to date are closely related to the skillfully designed coenzyme skeletons, as Jencks (1969) calls pyridoxal phosphate: the best catalyst God created. Finally, space precluded discussion of biotin, folic acid, ascorbic acid, vitamin B_{12}, vitamin K, etc. Ascorbic acid has frequently been employed as a reducing agent, but model studies of other coenzymes and vitamins are rather limited. To the eyes of organic chemists, the chemical functions of these coenzymes and vitamins are very fascinating as targets of biomimetic chemistry. It is to be expected that the further development of excellent coenzyme-originated catalysts will be continued.

REFERENCES

Abeles, R. H., Hutton, R. F. and Westheimer, F. H. (1957) *J. Am. Chem. Soc.*, **79**, 712.
Adams, M. J., Buehner, M., Chandrasekhar, K., Ford, G. C., Hockert, M. L., LiLjas, A., Rossman, M. G., Smiley, I. E., Allison, S., Evarse, J., Kaplan, N. O. and Taylor, S. (1973) *Proc. Natl. Acad. Sci. U.S.A.*, **70**, 1968.
Ball, S. and Bruice, T. C. (1980) *J. Am. Chem. Soc.*, **102**, 6498.
Bamberg, P. and Hemmerich, P. (1961) *Helv. Chim. Acta*, **44**, 1001.

Behr, J.-P. and Lehn, J.-M. (1978) *J. Chem. Soc., Chem. Commun.*, 143.
Belokon, Y. N., Tararov, V. I., Saveleva, T. F. and Belikov. V. M. (1980) *Makromol. Chem.*, **181**, 2183.
Breslow, R. (1958) *J. Am. Chem. Soc.*, **80**, 3719.
Breslow, R., Hammond, M. and Lauer, M. (1980) *J. Am. Chem. Soc.*, **102**, 421.
Brown, L. E. and Hamilton, G. A. (1970) *J. Am. Chem. Soc.*, **92**, 7225.
Bruice, T. C. (1980) *Acc. Chem. Res.*, **13**, 256.
Bruice, T. C. and Benkovic, S. J. (1966) *Bioorganic mechanisms*, Vol. 2, Benjamin, New York, Chapter 9.
Bruice, T. C., Chan, T. W., Taulane, J. P., Yokoe, I., Elliot, D. L., Williams, R. F. and Novak, M. (1977) *J. Am. Chem. Soc.*, **99**, 6713.
Buckley, T. F. and Rapoport, H. (1982) *J. Am. Chem. Soc.*, **104**, 4446.
Bunton, C. A., Minch, M. J., Hidalgo, J. and Sepulveda, L. (1973) *J. Am. Chem. Soc.*, **95**, 3262.
Bunton, C. A., Rivera, F. and Sepulveda, L. (1978) *J. Org. Chem.*, **43**, 1166.
Castells, J., Llitjos, H. and Moreno-Manas, M. (1977) *Tetrahedron Lett.*, 205.
Castells, J., Dunach, E., Geijo, F., Pujol, F. and Segura, P. M. (1978) *Isr. J. Chem.*, **17**, 278.
Cheeseman, P., Toms-Wood, A. and Wolfe, R. S. (1972) *J. Bacteriol.*, **112**, 527.
Christen, P. and Gasser, A. (1980) *Eur. J. Biochem.*, **107**, 73.
Clarke, M. J., Dowling, M. G., Garafalo, A. R. and Brennan, T. F. (1979) *J. Am. Chem. Soc.*, **101**, 223.
Cogoli-Greuter, M., Hausner, U. and Christen, P. (1979) *Eur. J. Biochem.*, **100**, 295.
Cookson, R. C. and Lane, R. M. (1976) *J. Chem. Soc., Chem. Commun.*, 804.
Corbett, M. D. and Chipko, B. R. (1980) *Bioorg. Chem.*, **9**, 273.
Creighton, D. J. and Sigman, D. S. (1971) *J. Am. Chem. Soc.*, **93**, 6314.
Crosby, J., Stone, R. and Lienhard, G. E. (1970) *J. Am. Chem. Soc.*, **92**, 2891.
Daigo, K. and Reed, L. J. (1962) *J. Am. Chem. Soc.*, **84**, 659.
Damen, J. and Neckers, D. C. (1980) *Tetrahedron Lett.*, 1913.
de Vries, J. G. and Kellog, R. M. (1979) *J. Am. Chem. Soc.*, **101**, 2759.
Dittmer, D. C. and Fouty, R. A. (1964) *J. Am. Chem. Soc.*, **86**, 91.
Douglas, K. T. and Nadvi, I. N. (1979) *FEBS Lett.*, **106**, 393.
Edmondson, D. E., Barman, B. and Tollin, G. (1972) *Biochemistry*, **11**, 1133.
Endo, T., Hayashi, Y. and Okawara, M. (1977) *Chem. Lett.*, 392.
Franzen, V. (1955) *Chem. Ber.*, **88**, 1361.
Franzen, V. and Fikentscher, L. (1958) *Liebigs. Ann. Chem.*, **613**, 1.
Fritchie, C. J., Jr. (1972) *J. Biol. Chem.*, **247**, 7459.
Frost, J. W. and Rastetter, W. H. (1981) *J. Am. Chem. Soc.*, **103**, 5242.
Gascoigne, I. M. and Radda, G. K. (1967) *Biochim. Biophys. Acta*, **131**, 498.
Gase, R. A. and Pandit, U. K. (1977) *J. Chem. Soc., Chem. Commun.*, 480.
Gase, R. A., Boxhoon, G. and Pandit, U. K. (1976) *Tetrahedron Lett.*, 2889.
Gibian, M. J. and Baumstark, A. L. (1971) *J. Org. Chem.*, **36**, 1389.
Gibian, M. J. and Rynd, J. A. (1969) *Biochem. Biphys. Res. Commun.*, **34**, 594.
Gibian, M. J. and Winkelman, D. W. (1969) *Tetrahedron Lett.*, 3901.
Gumbley, S. J. and Main, L. (1976) *Tetrahedron Lett.*, 3209.
Hadju, J. and Sigman, D. S. (1975) *J. Am. Chem. Soc.*, **97**, 3524.
Hadju, J. and Sigman, D. S. (1977) *Biochemistry*, **16**, 2841.
Hall, S. S., Doweyko, A. M. and Jordan, F. (1978) *J. Am. Chem. Soc.*, **100**, 5934.
Hastings, J. W., Balny, C., Le Peuch, C. and Douzou, P. (1973) *Proc. Natl. Acad. Sci. U.S.A.*, **70**, 3468.
Heggs, R. P. and Ganem, B. (1979) *J. Am. Chem. Soc.*, **101**, 2484.
Hemmerich, P., Nagelschneider, G. and Veeger, C. (1970) *FEBS Lett.*, **8**, 69.

Hemmerich, P. Massey, V. and Fenner, H. (1977) *FEBS Lett.*, **84**, 5.
Inoue, H. and Higashiura, K. (1980) *J. Chem. Soc., Chem. Commun.*, 549.
Inoue, H., Aoki, R. and Imoto, E. (1974) *Chem. Lett.*, 1157.
Iwata, M., Kuzuhara, H. and Emoto, S. (1976) *Chem. Lett.*, 983.
Jencks, W. P. (1969) *Catalysis in Chemistry and Enzymology*, McGraw-Hill, New York.
Jones, J. B. and Beck, J. F. (1976) in *Application of Biochemical Systems in Organic Chemistry* (eds J. B. Jones, C. J. Sih and D. Perlman), John Wiley & Sons, New York, pp. 107–401.
Kalyankar, G. D. and Snell, E. E. (1962) *Biochemistry*, **1**, 594.
Kemal, C. and Bruice, T. C. (1976a) *J. Am. Chem. Soc.*, **98**, 3955.
Kemal, C. and Bruice, T. C. (1976b) *Proc. Natl. Acad. Sci. U.S.A.*, **73**, 995.
Kemal, C. and Bruice, T. C. (1977) *J. Am. Chem. Soc.*, **99**, 7064.
Kemal, C. and Bruice, T. C. (1979) *J. Am. Chem. Soc.*, **101**, 1635.
Kemal, C., Chan, T. W. and Bruice, T. C. (1977) *J. Am. Chem. Soc.*, **99**, 7272.
Kemp, D. S. and Paul, K. G. (1970) *J. Am. Chem. Soc.*, **92**, 2553.
Kill, R. J. and Widdowson, D. A. (1976) *J. Chem. Soc., Chem. Commun.*, 755.
Kill, R. J. and Widdowson, D. A. (1978) in *Bioorganic Chemistry*, Vol. 4 (ed. E. E. van Tamelen), Academic Press, New York, pp. 239–275.
Kluger, R. and Pike, D. C. (1979) *J. Am. Chem. Soc.*, **101**, 6425.
Kluger, R., Chin, J. and Smyth, T. (1981) *J. Am. Chem. Soc.*, **103**, 884.
Knappe, W.-R. (1979) *Liebigs Ann. Chem.*, 1067.
Kojima, M., Toda, F. and Hattori, K. (1981) *J. Chem. Soc., Perkin Trans 1*, 1647.
Kosman, D. J. (1977) in *Bioorganic Chemistry*, Vol. 2 (ed. E. E. van Tamelen), Academic Press, New York, pp. 175–195.
Kozarich, J. W., Chari, R. V. J., Wu, J. C. and Lawrence, T. L. (1981) *J. Am. Chem. Soc.*, **103**, 4953.
Kunitake, T. and Shinkai, S. (1980) *Adv. Phys. Org. Chem.*, **17**, 435.
Kuzuhara, H., Iwata, M. and Emoto, S. (1977) *J. Am. Chem. Soc.*, **99**, 4173.
Kuzuhara, H., Komatsu, T. and Emoto, S. (1978) *Tetrahedron Lett.*, 3563.
Llor, J. and Cortijo, M. (1977) *J. Chem. Soc., Perkin Trans 2*, 1111.
Loechler, E. L. and Hollocher, T. C. (1975) *J. Am. Chem. Soc.*, **97**, 3235.
MacInnes, I., Nenhebel, D. C., Orszulik, S. T. and Suckling, C. J. (1982) *J. Chem. Soc., Chem. Commun.*, **121**, 1146.
Main, L., Dasperek, G. J. and Bruice, T. C. (1972) *Biochemistry*, **11**, 3991.
Makino, T., Nunozawa, T., Baba, N. Oda, J. and Inoue, Y. (1979) *Tetrahedron Lett.*, 1683.
McCapra, F. and Lesson, P. (1976) *J. Chem. Soc., Chem. Commun.*, 1037.
Metzler, D. E., Ikawa, M. and Snell, E. E. (1954) *J. Am. Chem. Soc.*, **76**, 648.
Miller, A. (1982) *Tetrahedron Lett.*, **23**, 753.
Moras, D., Olsen, K. W., Sabesan, M. N., Buehner, M., Ford, G. C. and Rossman, M. G. (1975) *J. Biol. Chem.*, **250**, 9137.
Murakami, Y. and Kondo, H. (1975) *Bull. Chem. Soc. Jpn*, **48**, 541.
Murakami, Y., Aoyama, Y. and Kikuchi, J. (1981a) *J. Chem. Soc., Chem. Commun.*, 444.
Murakami, Y., Aoyama, Y. and Kikuchi, J. (1981b) *J. Chem. Soc., Perkin Trans, 1*, 2809.
Muto, S. and Bruice, T. C. (1980) *J. Am. Chem. Soc.*, **102**, 7559.
Nakano, H., Nishioka, M., Sangen, O. and Yamamoto, Y. (1981) *J. Polym. Sci., Polym. Chem. Ed.*, **19**, 2919.
Nambu, Y., Endo, T. and Okawara, M. (1980) *J. Polym. Sci., Polym. Chem. Ed.*, **18**, 2793.

Nambu, Y., Endo, T. and Okawara, M. (1981) *J. Polym. Sci., Polym. Chem. Ed.*, **19**, 1937.

Nishide, H., Deguchi, J. and Tsuchida, E. (1976) *Chem. Lett.*, 169.

Nishiyama, K., Baba, N., Oda, J. and Inoue, Y. (1976) *J. Chem. Soc., Chem. Commun.*, 101.

Oae, S., Tagaki, W. and Ohno, A. (1961) *J. Am. Chem. Soc.*, **83**, 5036.

O'Brien, D. E., Weinstock, L. T. and Cheng, C. C. (1970) *J. Heterocyclic Chem.*, **7**, 99.

Ohnishi, Y. and Kitami, M. (1978) *Tetrahedron Lett.*, 4035.

Ohnishi, Y., Kagami, M. and Ohno, A. (1975a) *Tetrahedron Lett.*, 2437.

Ohnishi, Y., Kagami, M. and Ohno, A. (1975b) *Chem. Lett.*, 125.

Ohnishi, Y., Kagami, M. and Ohno, A. (1975c) *J. Am. Chem. Soc.*, **97**, 4766.

Ohnishi, Y., Kagami, M., Numakunai, T. and Ohno, A. (1976) *Chem. Lett.*, 915.

Ohno, A., Ikeguchi, M., Kimura, T. and Oka, S. (1979) *J. Am. Chem. Soc.*, **101**, 7063.

Ohno, A., Yasui, S., Gase, R. A., Oka, S. and Pandit, U. K. (1980) *Bioorg. Chem.*, **9**, 199.

Ono, N., Tamura, R. and Kaji, A. (1980) *J. Am. Chem. Soc.*, **102**, 2851.

Ono, N., Tamura, R., Tanikaga, R. and Kaji, A. (1981) *J. Chem. Soc., Chem. Commun.*, 71.

Pac, C., Ihama, M., Yasuda, M., Miyauchi, Y. and Sakurai, H. (1981) *J. Am. Chem. Soc.*, **103**, 5495.

Pandit, U. K. and Mas Cabré, F. R. (1971) *J. Chem. Soc., Chem. Commun.*, 552.

Pattison, S. E. and Dunn, M. F. (1976) *Biochemistry*, **15**, 3691.

Racker, E. (1951) *J. Biol. Chem.*, **190**, 685.

Rastetter, W. H. and Adams, J. (1981) *J. Org. Chem.*, **46**, 1882.

Rastetter, W. H., Gadek, T. R., Tane, J. P. and Frost, J. W. (1979) *J. Am. Chem. Soc.*, **101**, 2228.

Rynd, J. A. and Gibian, M. J. (1970) *Biochem. Biophys. Res. Commun.*, **41**, 1097.

Seki, M., Baba, N., Oda, J. and Inoue, Y. (1981) *J. Am. Chem. Soc.*, **103**, 4613.

Selbin, J., Sherrill, J. and Bigger, C. H. (1974) *Inorg. Chem.*, **13**, 2544.

Shea, K. J., Thompson, E. A., Pandey, S. D. and Beauchamp, P. S. (1980) *J. Am. Chem. Soc.*, **102**, 3149.

Sheehan, J. C. and Hara, T. (1974) *J. Org. Chem.*, **39**, 1196.

Shinkai, S. (1982) *Prog. Polym. Sci.*, **8**, 1.

Shinkai, S. and Bruice, T. C. (1973a) *Biochemistry*, **12**, 1750.

Shinkai, S. and Bruice, T. C. (1973b) *J. Am. Chem. Soc.*, **95**, 7526.

Shinkai, S. and Kunitake, T. (1977) *Chem. Lett.*, 297.

Shinkai, S., Kunitake, T. and Bruice, T. C. (1974) *J. Am. Chem. Soc.*, **96**, 7140.

Shinkai, S., Ando, R. and Kunitake, T. (1975) *Bull. Chem. Soc. Jpn*, **48**, 1914.

Shinkai, S., Shiraishi, S. and Kunitake, T. (1976a) *Bull. Chem. Soc. Jpn*, **49**, 3656.

Shinkai, S., Ando, R. and Kunitake, T. (1976b) *Bull. Chem. Soc. Jpn*, **49**, 3652.

Shinkai, S., Hamada, H., Ide, T. and Manabe, O. (1978a) *Chem. Lett.*, 685.

Shinkai, S., Ide, T. and Manabe, O. (1978b) *Bull. Chem. Soc. Jpn*, **51**, 3655.

Shinkai, S., Yamada , S. and Kunitake, T. (1978c) *J. Polym. Sci., Polym. Lett. Ed.*, **16**, 137.

Shinkai, S., Ando, R. and Kunitake, T. (1978d) *Biopolymers*, **17**, 2757.

Shinkai, S., Yamada, S. and Kunitake, T. (1978e) *Macromolecules*, **11**, 65.

Shinkai, S., Hamada, H., Kusano, Y. and Manabe, O. (1979a) *Bull. Chem. Soc. Jpn*, **52**, 2600.

Shinkai, S., Hamada, H., Kusano, Y. and Manabe, O. (1979b) *J. Chem. Soc., Perkin Trans, 2*, 699.

Shinkai, S., Hamada, H. and Manabe, O. (1979c) *Tetrahedron Lett.*, 1397.
Shinkai, S., Hamada, H., Kusano, Y. and Manabe, O. (1979d) *Tetrahedron Lett.*, 3511.
Shinkai, S., Nakano, T., Hamada, H., Kusano, Y., and Manabe, O. (1979e) *Chem. Lett.*, 229.
Shinkai, S., Mori, K., Kusano, Y. and Manabe, O. (1979f) *Bull. Chem. Soc. Jpn*, **52**, 3606.
Shinkai, S., Hamada, H., Dohyama, A. and Manabe, O. (1980a) *Tetrahedron Lett.*, 1661.
Shinkai, S., Kusano, Y., Manabe, O. and Yoneda, F. (1980b) *J. Chem. Soc., Perkin Trans 2*, 1111.
Shinkai, S., Yamashita, T., Kusano, Y., Ide, T. and Manabe, O. (1980c) *J. Am. Chem. Soc.*, **102**, 2335.
Shinkai, S., Yamashita, T., Kusano, Y. and Manabe, O. (1980d) *Tetrahedron Lett.*, **21**, 2543.
Shinkai, S., Yamashita, T., Kusano, Y. and Manabe, O. (1980e) *J. Org. Chem.*, **45**, 4947.
Shinkai, S., Hamada, H., Kuroda, H. and Manabe, O. (1981a) *J. Org. Chem.*, **46**, 2333.
Shinkai, S., Yamashita, T., Kusano, Y. and Manabe, O. (1981b) *J. Am. Chem. Soc.*, **103**, 2070.
Shinkai, S., Ishikawa, Y. and Manabe, O. (1982a) *Chem. Lett.*, 809.
Shinkai, S., Yamashita, T., Kusano, Y. and Manabe, O. (1982b) *J. Am. Chem. Soc.*, **104**, 563.
Shinkai, S., Hara, Y. and Manabe, O. (1982c) *J. Polym. Sci., Polym. Chem. Ed.*, **20**, 1097.
Shinkai, S., Ishikawa, Y. and Manabe, O. (1983) *Bull. Chem. Soc. Jpn*, **56**, 1964.
Shirra, A. and Suckling, C. J. (1977) *J. Chem. Soc., Perkin Trans, 2*, 759.
Sigman, D. S., Hadju, J. and Creighton, D. J. (1978) in *Bioorganic Chemistry*, Vol. 4 (ed. E. E. van Tamelen), Academic Press, New York, pp. 385–407.
Spencer, R., Fisher, J. and Walsh, C. (1977) *Biochemistry*, **16**, 3586.
Spetnagel, W. J. and Klotz, I. M. (1978) *Biopolymers*, **17**, 1657.
Steffens, J. J. and Chipman, D. M. (1971) *J. Am. Chem. Soc.*, **93**, 6694.
Stetter, H. and Kuhlmann (1976) *Chem. Ber.*, **109**, 2890.
Stetter, H., Basse, W. and Nienhaus, J. (1980) *Chem. Ber.*, **113**, 690.
Tagaki, W. and Hara, H. (1973) *J. Chem. Soc., Chem. Commun.*, 891.
Tagaki, W., Tamura, Y. and Yano, Y. (1980) *Bull. Chem. Soc. Jpn*, **53**, 478.
Takagi, M. Goto, S., Ishihara, R. and Matsuda, T. (1976) *J. Chem. Soc., Chem. Commun.*, 993.
Takagi, M., Goto, S., Tazaki, M. and Matsuda, T. (1980) *Bull. Chem. Soc. Jpn*, **53**, 1982.
van Bergen, T. J. and Kellog, R. M. (1977) *J. Am. Chem. Soc.*, **99**, 3882.
van Eikeren, P. and Grier, D. L. (1976) *J. Am. Chem. Soc.*, **98**, 4655.
van Ramesdonk, H. J., Verhoeven, J. W., Pandit, U. K. and de Boer, Th. J. (1978) *Rec. Trav. Chim. Pays-Bas.*, **97**, 195.
Wallenfels, K. and Hanstein, W. (1965) *Angew. Chem. Int. Ed. Engl.*, 869.
Wallenfels, K., Ertel, W. and Friedrich, K. (1973) *Liebigs Ann. Chem.*, 1663.
Walsh, C. (1980) *Acc. Chem. Res.*, **13**, 148.
Walsh, C., Schonbrunn, A. and Abeles, R. H. (1971) *J. Biol. Chem.*, **246**, 6855.
Walsh, C., Fisher, J., Spencer, R., Graham, D. W., Ashton, W. T., Brown, J. E., Brown, R. D. and Rogers, E. F. (1978) *Biochemistry*, **17**, 1942.
White, F. G. and Ingraham, L. L. (1962) *J. Am. Chem. Soc.*, **84**, 3109.

Williams, R. F. and Bruice, T. C. (1976) *J. Am. Chem. Soc.*, **98**, 7752.

Wilson, G. E., Jr. and Hess, A. (1980) *J. Org. Chem.*, **45**, 2766.

Wulff, G., Vesper, W., Grode-Einsler, R. and Sarhan, A. (1977) *Makromol. Chem.*, **178**, 2799.

Yaggi, N. F. and Douglas, K. T. (1977) *J. Am. Chem. Soc.*, **99**, 4844.

Yano, Y., Hoshino, Y. and Tagaki, W. (1980) *Chem. Lett.*, 749,

Yokoe, I. and Bruice, T. C. (1975) *J. Am. Chem. Soc.*, **97**, 450.

Yoneda, F. and Nakagawa, K. (1980) *J. Chem. Soc., Chem. Commun.*, 878.

Yoneda, F., Sakuma, Y. and Hemmerich, P. (1977) *J. Chem. Soc., Chem. Commun.*, 825.

Yoneda, F., Kawazoe, M. and Sakuma, Y. (1978a) *Tetrahedron Lett.*, 2803.

Yoneda, F., Sakuma, Y. and Nitta, Y. (1978b) *Chem. Lett.*, 1177.

Yoneda, F., Hirayama, R. and Yamashita, M. (1980a) *Chem. Lett.*, 1157.

Yoneda, F., Ono, M., Kira, K., Tanaka, H., Sakuma, Y. and Koshiro, A. (1980b) *Chem. Lett.*, 817.

Yoneda, F., Yamato, H. and Ono, M. (1981) *J. Am. Chem. Soc.*, **103**, 5943.

4 | Selectivity in synthesis – chemicals or enzymes

Colin J. Suckling

4.1 INTRODUCTION

A characteristic feature of recent developments in organic synthesis is an emphasis upon selectivity. In chemical reactions, selectivity takes many forms. At the simplest level, the chemist is concerned with ensuring that, as far as possible, only the desired reaction takes place (chemoselectivity), and one of the first areas to reach a high state of development in this context was peptide synthesis. The selective synthesis of peptides has been possible because protecting or blocking groups were devised to permit the sequential construction of the required peptide. Today, an experienced peptide chemist can make near-optimal choices of protecting groups. Further developments of protecting-group chemistry have led to efficient procedures for the specific synthesis of oligonucleotides.

At the next level of control, the chemist wishes to determine at which of two or more positions of similar reactivity a reaction will occur (regioselectivity). There are venerable chemical strategies for controlling regioselectivity. For instance, it has usually been found that the less reactive a reagent, the more selective its action; also, it is sometimes possible to control a reaction by choosing the conditions such that either the kinetically or the thermodynamically favoured product predominates. An example of the former is the greater selectivity of bromine atoms in alkane halogenation compared with the more reactive chlorine, and the latter may be exemplified by the regiospecific formation of ketone enolates according to the reaction conditions. Perhaps the most subtle control required of the organic chemist is the control of stereochemistry and in particular the formation at will of only one optical isomer (stereoselectivity). In recent years, the control of stereo-chemistry has taken pride of place in most synthetic investigations. As we shall see, enzyme chemistry can contribute greatly both directly and in-directly to control in synthesis. For convenience in the discussion, I shall refer to the use of enzymes and micro-organisms in synthesis as biochemical methods and conventional reactions involving neither as chemical methods.

There are several reasons for the modern chemist's preoccupation with selectivity. He has the academic aim of controlling the regio- and stereo-selectivity of a synthetic reaction as completely as possible; this endeavour has its origins not only in striving for elegance in the execution of a synthesis, but also in practical reasons. On the one hand, many naturally occurring compounds, which are common synthetic targets, are biologically active in only one stereoisomeric form and on the other hand, the economic pressure in industry to avoid wasteful and possibly harmful by-products daily demands more attention from the chemist.

The word 'selectivity' features in the titles of many current papers and in a recent article Norman (1982) identified several strategies open to the chemist to control selectivity. He considered established chemical devices such as neighbouring-group effects and the reactivity-controlling methods men-tioned above but added for the first time in such a broad discussion the challenge of enzymic selectivity by including some biomimetic examples. However, with the exception of polynucleotide synthesis, chemists have been slow to make use of enzymes. It is probably not generally realized that the number of synthetic reactions to which enzyme chemistry can be applied is very large. Perhaps it is often difficult to see how the use of an enzyme might aid a synthesis or at what stage in a synthesis enzyme chemistry can be of most value. This chapter explores these avenues by comparing the current ability of selected chemical reagents to control synthetic transformations with the properties of closely analogous enzyme-catalysed reactions. With an eye to the future, the achievements and potential of biomimetic systems in synthesis will also be discussed.

As I have said, until about ten years ago, very little attention had been paid to the harnessing of enzymes in general organic synthesis. An illustration of the prevailing view at that time is provided by a few lines from Fleming's valuable teaching book *Selected Organic Syntheses* (1973).

'Actually, this . . . synthesis, using a microbiological oxidation, is not a total synthesis, even in principle; the use of microorganisms which have not themselves been synthesised disqualifies the route as a total synthesis in the usually accepted academic sense.'

And later

'They (enzymes) have the inestimable advantage of being able to . . . carry out highly specific reactions, whether in a polyfunctional molecule or in a completely unfunctionalised one. . . . They are not convenient laboratory reagents, although some enzymes, such as ribonuclease, which have rather predictable properties certainly are.'

These lines are suggestive of a rigorous academic viewpoint but they highlight accurately the situation in the early 1970s. Interestingly, only two examples of biochemical synthetic steps appear in the book and both refer to industrial syntheses. Fleming's comment 'From the point of view of industrial practice, this was simply the best way to do the reaction' is surely sound advice to any modern synthetic chemist interested in reaching a target. Nevertheless, even ten years ago the methodology for using enzymes and micro-organisms was not sufficiently well advanced to attract the attention of synthetic chemists at large. The potential was clearly there, but so were the problems (Suckling and Suckling, 1974). Principal amongst these were the stability of enzymes in use and, most significantly, the recycling of co-enzymes. The nature of coenzymes as essentially enzyme-bound reagents will have become obvious from the preceding chapter. Such great strides have been made in solving these problems, especially in the laboratories of Jones at Toronto and Whitesides at MIT (Whitesides, 1976), that it is now possible to compare enzymes and conventional procedures on equal terms. However, the third arm of this chapter's comparison, biomimetic chemistry, is currently as enzymes were ten years ago, promising but problematical.

It is important to distinguish between the use of purified enzyme preparations, which contain essentially only one catalytic activity, and microbiological fermentations, which contain many. Obviously a synthetic chemist requires predictability in his reactions, and a microbiological method will not usually be attractive although, as we shall see, it can be most effective. On the other hand, many enzyme-catalysed reactions are today highly predictable and have earned their place in the synthetic organic chemist's repertoire. Regrettably, most standard reference works to organic synthesis omit enzymes. Fieser's Reagents series lists none, but Theilheimer is a useful source. The most extensive compilation of relevant information is in Weissberger's *Techniques of Organic Chemistry Series* (Jones *et al.*, 1976). In view of this situation, it might pay one of the enzyme suppliers to promote a commercial catalogue of synthetically useful enzymes. Their sales slogan might be 'An accessible source of controlled selectivity'.

4.2 PROBLEMS OVERCOME

To capitalize upon the selective advantages of enzymes, it was essential that the practical problems of coenzyme recycling and of enzyme stability be solved. A good example to illustrate developments of the last decade in the use of

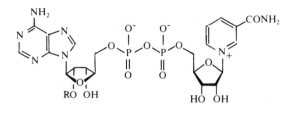

R=H NAD$^+$ R=PO$_3$H$_2$ NADP$^+$

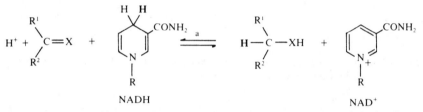

NADH   NAD$^+$

Fig. 4.1 Nicotinamide–adenine dinucleotide and some reactions mediated by dehydrogenases, e.g. a = horse liver alcohol dehydrogenase (HLADH), R^1 = alkyl, R^2 = H, X = O; lactate dehydrogenase (LDH), R^1 = CH$_3$, R^2 = CO$_2$H, X = O.

enzymes in synthesis is the recycling of the redox coenzyme, nicotinamide–adenine dinucleotide (NAD) or its 3'-phosphate (NADP). It has often been argued that the carbonyl group is the most important functionality in synthesis. The ability to reduce aldehydes and ketones to alcohols with control of chirality of the product is obviously significant. The conventional chemical solution is to use bulky groups either in the substrate or the reagent to hinder approach from one of the two enantiotropic faces of the carbonyl group. An enzyme has a structure that will control the direction of attack. However, such redox enzymes (dehydrogenases) require the assistance of the coenzyme NAD in its reduced form, NADH, and in stoichiometric quantities (Fig. 4.1). This coenzyme is much more costly (£10 000 per mole) than the most expensive competitive selective chemical reducing agent. It is therefore essential that they be recycled if the enzyme-catalysed reduction of carbonyl groups is to be practical in synthesis. The first extensively used method employed *coupled substrates* (e.g. Battersby *et al.*, 1975) and illustrates elegantly a further ramification of enzymes in synthesis, reversible catalysis (Fig. 4.2). In this reduction of an aldehyde, an excess of an alcohol was used: the oxidized NAD would then act with the enzyme to oxidize the co-substrate alcohol and thereby regenerate the required reducing agent, NADH. This method has been widely applied in the synthesis of chirally

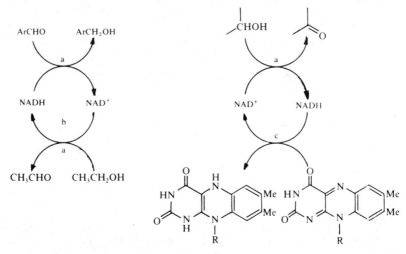

Fig. 4.2 Some recycling methods for NAD⁺/NADH. a, HLADH; b, Na₂S₂O₄; c, chemical oxidation using flavin mononucleotide.

labelled benzylic alcohols. Shortly afterwards, Jones and Beck (1976) and Jones and Taylor (1976) showed that coenzyme recycling could also be accomplished using added chemical oxidizing or reducing agents. If NAD(P)H was required for a reduction, sodium dithionite served to regenerate the coenzyme and, if an oxidation was needed, flavin mononucleotide (FMN) was effective in recycling NAD(P).

However, none of these methods was applied on quantities larger than a few grams of substrate, and it was left to Whitesides to give a virtuoso exposition of recycling methods for NAD(P)H on up to a mole scale. The success of Whitesides' work is based upon the fact that it is possible to use a mixture of several enzymes, each with different catalytic activities, in the same reaction flask without one interfering with the other. Chemical reagents are scarcely ever so compatible. Using a *coupled enzyme*, NAD(P)⁺ can be reduced to NAD(P)H with concomitant oxidation of a substrate for the second enzyme. Experience has shown that the best enzyme for relaying the reduction of NAD(P)⁺ is glucose 6-phosphate dehydrogenase; thus glucose 6-phosphate becomes the terminal reductant or fuel for the reduction (Fig. 4.3). However, since glucose 6-phosphate is not cheap, it would be advantageous if it too could be regenerated enzymically. This can be achieved using hexokinase, the enzyme that catalyses the phosphorylation of glucose by ATP. The enzyme chain has now relayed the task of driving the whole reaction sequence to ATP which can fortunately be produced cheaply either chemically or enzymically (Lewis *et al.* 1979; Leuchs *et al.*, 1979).

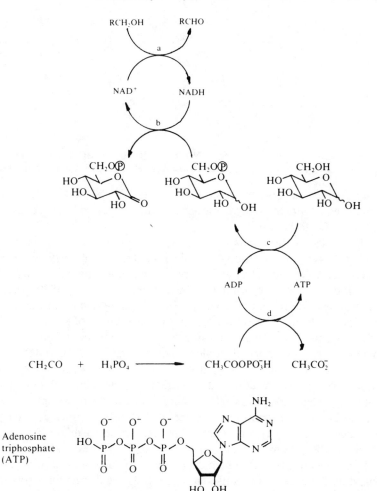

CH₂CO + H₃PO₄ ⟶ CH₃COOPO₃H   CH₃CO₂⁻

Adenosine
triphosphate
(ATP)

Fig. 4.3 ATP and NAD⁺ recycling coupled together. a, HLADH; b, glucose 6-phosphate dehydrogenase; c, hexokinase; d, acetate kinase.

Such chains of coupled enzyme-catalysed reactions are efficient when used on up to a few moles of reactants but on larger scales still, the number and quantity of co-operating enzymes and relay substrates become prohibitively large. It then becomes appropriate to use electrochemical methods or hydrogen gas as the terminal reductant. In the former case, a coupled system is still used in which Methyl Viologen acts as an electron carrier from the cathode to the enzyme that reduces NAD(P)⁺ (Fig. 4.4; DiCosimo *et al.*, 1981). Such procedures are not restricted to purified enzymes as Simon and his collegues have shown. Reductions by micro-organisms can also be driven

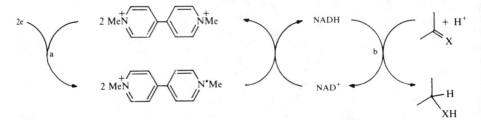

Fig. 4.4 Electrochemical recycling of NADH. a, Reduction of Methyl Viologen at a cathode (W or Hg); b, an enzymic reduction, e.g. HLADH or enoate reductase.

by electrochemical energy (Simon *et al.*, 1981). When hydrogen is the terminal reductant, the relay component is an enzyme, a hydrogenase, that reduces NAD^+ to NADH.

At this stage in the discussion, it is worth making the first direct comparison between a conventional chemical reagent and an enzyme-mediated synthetic step. The reduction of deuteriated benzaldehydes has been particularly well documented for both approaches. Midland *et al.* (1979) described the synthesis of chiral deuteriobenzyl alcohols using chiral boranes derived from pinene. His group achieved an enantiomeric excess of greater than 90% for the S-isomer on a small laboratory scale. The analogous enzymic reaction was catalysed by horse liver alcohol dehydrogenase (HLADH) and was carried out on 0.4 mole (Wong and Whitesides, 1981). A product with 95% enantiomeric excess was obtained. Significantly, the same technology using lactate dehydrogenase was used to prepare D-lactate in similar optical purity. It cannot be doubted from these examples that both chemical and enzymic techniques are practical. In Fleming's terms, the purist might discount the enzymic method from the point of view of total synthesis. However, since the source of chirality in the chemical procedure is pinene, a natural product of enzymic asymmetric induction in a tree, the only distinction between the two methods is the stage at which the enzyme acts. Objections on philosophical grounds to the use of enzymes in synthesis therefore appear to be weakly founded. It's not cheating to use an enzyme in synthesis; if it does the job, it's useful, if it doesn't, try something else.

In his preparative reductions just described, Whitesides made extensive use of *immobilized enzymes*. Just as immobilized chemical reagents offer advantages in ease of work-up, so immobilized enzymes permit the rapid separation of the expensive catalyst from the solution of reactants and products (Chibata, 1978; Suckling, 1977). This is valuable but more importantly, it has been generally found that immobilization enhances the stability of the enzyme. In this way, it is possible to solve the second practical problem of enzymes in synthesis. Of the many methods available for immobilizing an enzyme, a specially synthesized polymer of acrylamide and the N-hydroxysuccinimide ester of acrylic acid has proved widely applicable (Pollak *et al.*, 1977, 1978). The risk of damaging the enzyme's active site through attack by the immobilizing reagents can be minimized by carrying out immobilization in the presence of the substrate or an inhibitor, either of which will bind at the active site and protect it from attack. In this way, a remarkably large number of enzymes has been immobilized, many in quantities sufficient for chemical synthesis on a mole scale. In industrial applications, the ability to obtain an active, stable enzyme preparation makes the difference between viability and otherwise of a process. There are several clear examples of this including the use of columns of immobilized fungal cells to catalyse specifically 11-β-hydroxylation of steroids and a subsequent 1,2-dehydrogenation (Mosbach and Larsson, 1970; Chibata, 1978) leading to corticosteroids: these reactions are related to those described by Fleming (1973). Notably, the microbial transformations using the immobilized catalyst carry out firstly a regio- and stereo-selective hydroxylation at a non-activated carbon atom and secondly a specific dehydrogenation. Both of these transformations are impossible using conventional chemical reagents. Later in this chapter, however, we shall see how biomimetic experiments can approach such problems.

A further aspect of enzyme stability is the sensitivity of many enzymes to oxygen. Usually such enzymes contain important thiol groups that are readily oxidized by air to catalytically inactive disulphides. To prevent oxidative inactivation, the enzyme-catalysed reaction can be carried out under an inert atmosphere, preferably argon, in the presence of an added thiol such as dithiothreitol or mercaptoethanol; these thiols are oxidized by adventitious oxygen in place of the enzyme's sensitive functional group. With characteristic thoroughness, Whitesides' group has also studied this problem (Szajewski and Whitesides, 1980). The experimental technique is no more difficult than handling modern air-sensitive reagents. A further extension to the general applicability of enzymes in organic synthesis is shown by the ability of enzymes to tolerate organic solvents. For example, Jones and Schwartz (1982) have recently shown that HLADH can function with stereospecificity in the presence of up to 50% of many water-miscible solvents.

4.3 LOGIC AND ANALOGY IN THE SYNTHETIC USES OF ENZYMES AND MICRO-ORGANISMS

Every modern organic chemist learns his craft with the aid of structural and mechanistic analogy. If biochemical methods are to be readily acceptable, it is not unreasonable for the chemist to expect to be able to incorporate enzyme-catalysed reactions into the same system. Even micro-organisms show recognizable reactivity patterns towards organic compounds in the view of Sih and Rosazza (1976) and Perlman (1976). They point out that yeasts readily transform carbonyl groups, fungi cause hydroxylations, but bacteria are generally too destructive. Aerobic micro-organisms are especially simple to use in synthesis because, unlike purified enzymes, they can recycle their own coenzymes; all they need is a growth medium containing suitable nutrients. The major practical problem, which can be serious on a large scale, is the isolation of the product. However attractive microbiological methods may appear in their simplicity, most chemists will not show interest unless telling examples are available. Fortunately, some recent work using yeast to prepare chiral synthons as the basis of complex natural product syntheses have been described (Fronza *et al.*, 1980; Fuganti and Graselli, 1979, 1982). It has been known for many years that yeast will reduce ketones to chiral secondary alcohols; the reaction is even in *Organic Syntheses*. Fuganti has ingeniously extended this utility by capitalizing upon the observation that cinnamaldehydes, and homologues, undergo not only carbonyl group reduction, but also C=C reduction and aldol condensation (Fig. 4.5). The benzylic fragment is redundant after it has channelled the substrate into yeast metabolism; as a synthon, it acts as a latent carbonyl function. The target molecules were important natural products, namely tocopherol, a natural anti-oxidant found in lipids, and daunosamine, a component of the anti-tumour antibiotic daunomycin. The repeated chiral structures of tocopherol's side chain make this synthetic strategy ideal. The ready availability of yeast and the clear opportunity for logical extension of these reactions should attract further synthetic applications. For comparison with chemical methods, the use of the stereospecific addition of carbanions to vinyl lactones controlled by palladium(0) complexes as templates illustrates an alternative strategy for the construction of the tocopherol side chain (Trost and Klun, 1979, 1981). Template strategies figure in biomimetic chemistry discussed below.

Many enzymes have been so thoroughly studied that their behaviour in synthesis can be predicted reliably. These include the peptidase, chymotrypsin, for which an active-site model has been deduced from a combination of specificity studies and X-ray crystallography on the enzyme (Jones and Beck, 1976). Undoubtedly the classic example of the development of predictability concerns the enzyme horse liver alcohol dehydrogenase, which has been mentioned above (Fig. 4.1). The early studies of this enzyme in the laboratories of Prelog in Zurich have recently been spectacularly extended by Jones' group in Toronto (Jones, 1980). HLADH is a relatively inexpensive enzyme. Although specific with regard to the chirality of the redox reaction it catalyses at the carbonyl group of the substrate, it will accept an unusually wide range of organic substrates containing typically 2 to 15 carbon atoms. The broad specificity of this enzyme probably reflects the biological source, liver, which is the organ principally responsible for detoxification of organic compounds foreign to the metabolism of the

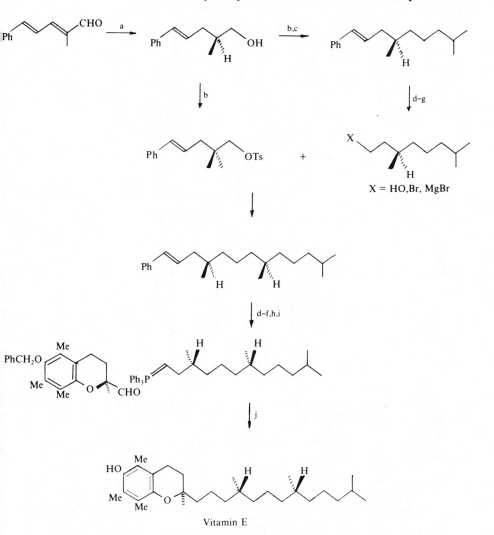

Fig. 4.5 Multiple use of a yeast-generated chiral building block in a synthesis of vitamin E. a, Fermenting yeast; b, TsCl/py; c, BrMg(CH₂)CHMe₂ LiCuCl₄; d, O₃, e, LiAlH₄; f, NBS Ph₃P; g, Mg; h, Ph₃P; i, base; j, H₂/cat.

animal. It may well be that other synthetically useful enzymes are present in the same source. The structural origin of the broad specificity of HLADH has been identified as a large non-polar pocket adjacent to the catalytically active site; this binding site is especially suited to alicyclic molecules. What the synthetic chemist wants to know is whether his substrate will fit the binding site and will be reduced at a useful rate by HLADH. Prelog sought an

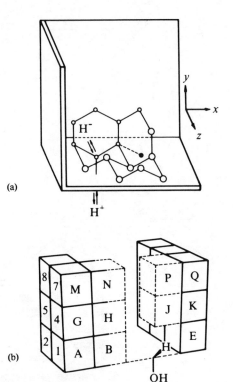

Fig. 4.6 (a) Prelog's diamond lattice model. (b) Jones' cubic model. Jones' model is depicted viewed from a left-front viewpoint. The forbidden regions are shown bounded by solid (——) lines and the limited regions by broken (– – –) lines. The open spaces are the allowed areas favouring substrate binding. For specificity analyses, the substrate model is built to the same scale. Its C=O or CH(OH) group is first positioned in the required orientation at the oxidation site located at the bottom of the Cl,Dl intersection. Its various orientations in the cubic lattice section are then compared in order to identify which, if any, will permit the formation of a productive ES complex. Both figures reproduced with permission; Fig. 4.6(a) by courtesy of *Pure Appl. Chem.*, IUPAC.

answer to this question by offering the enzyme a series of substituted cyclo-hexanones and decalinones. Some compounds were found to be reduced rapidly, some slowly, and some not at all. It was then possible to assess where substituents favoured binding and reduction, and where they hindered it. Since cycloalkanones are chiefly built from tetrahedral carbon atoms, Prelog constructed a reactivity map in the form of a so-called 'diamond lattice' (Fig. 4.6(a); Prelog, 1964). The chemist could then fit his substrate to the lattice and see whether any unfavourable interactions were present.

Undoubtedly, Prelog's work established the predictability of enzyme-catalysed reactions in synthesis and it was natural that when beginning his

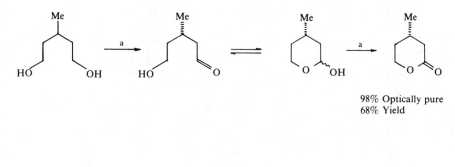

98% Optically pure
68% Yield

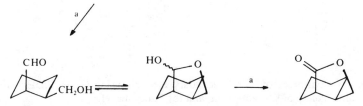

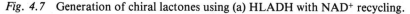

Fig. 4.7   Generation of chiral lactones using (a) HLADH with NAD⁺ recycling.

studies, Jones should build upon Prelog's foundation. At that time, the only recognized substrates for HLADH were ethanol, benzaldehyde and substituted cyclic alkanones. There were some notable gaps in possible substrates. For instance, aldehydes were easily reduced by the enzyme; if labelled substrate or coenzyme was used, a chiral product resulted. However, acyclic ketones were poor substrates and also, no attention had been paid to stereochemical control in oxidative reactions. The newly discovered NAD recycling technique made it possible for Jones to attempt oxidations of achiral compounds and use the enzyme's natural asymmetry to induce chirality in the product (Jones and Lok, 1979). Fig. 4.7 illustrates the oxidation of achiral alcohols to afford good yields of optically pure lactones. Two enzyme-catalysed reactions occur: firstly, the enzyme selects one arm of the substrate for oxidation, the pro-*R* in this case, and secondly, the hemiacetal which is in equilibrium with the aldol is oxidized to the lactone.

In order to convert an acyclic ketone into a substrate, Jones constrained the ketone into a ring using sulphur which could readily be removed with Raney nickel after chiral reduction. Thus the synthetic equivalent of an acyclic ketone became, from the enzyme's point of view, a welcome

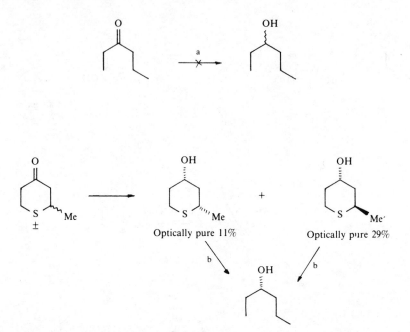

Fig. 4.8 Syntheses of chiral secondary alcohols using (a) HLADH with NAD⁺ recycling (run to 50% completion) and (b) Raney nickel.

cycloalkyl substrate (Fig. 4.8; Jones and Davies, 1979; Jones and Schwartz, 1981).

The extended utility of HLADH illustrated by the above examples nevertheless exposed a limitation in the predictive value of the diamond lattice model. Five-membered rings and compounds containing sulphur do not readily fit the points of the diamond lattice. Clearly what was required was a model that specified regions where substituents could be tolerated and where they could not. Jones has constructed a convenient model based upon cubic spatial sections which allows the chemist to assess easily the likely success of an oxidation or reduction (Fig. 4.6(b); Jones and Jakovac, 1982).

The current understanding of alcohol dehydrogenase in synthesis approaches an ideal state of the art. Both the scope and limitations of its efficacy in preparing chiral synthons are clearly and accessibly established so that a new reaction may be approached with confidence. Many new chemical reagents are less well documented. It is to be hoped that in the future, more enzymes will be similarly evaluated.

4.4 ENZYMES AND CHEMICAL REAGENTS IN 'COMPETITION'

From the foregoing discussion there can be no doubt that biochemical synthetic methods are amenable to the normal logical processes of organic chemistry. It is now appropriate to examine how biochemical methods fare in competition with chemical reagents in the tough testing ground of natural product synthesis and the related synthesis of labelled precursors for biosynthetic studies. There are three areas to consider, synthon preparation and functional group modification, but firstly I shall discuss the biosynthetically important area of chiral isotopically labelled compounds.

4.4.1 Chiral isotopic labels

The synthesis of chiral isotopically labelled compounds was one of the first fields in which the significance of enzyme stereoselectivity was realized. Indeed the important initial demonstration of enzymic selectivity at pro-chiral centres demanded the use of isotopic labels. One example, the reduction of benzaldehyde, has already been described. Related reactions with substituted benzaldehydes have been especially valuable in the synthesis of precursors for biosynthetic studies of alkaloids (Battersby and Staunton, 1974). The use of tritiated and deuteriated phenethylamines has permitted the elucidation of the stereochemical course of benzylic hydroxylation in alkaloid biosynthesis. However, a less frequently discussed example concerns the synthesis of $[^{13}C]$-valine for the study of the biosynthesis of β-lactam antibiotics, a problem considered from a different perspective in Chapter 7.

There are two nicely contrasting syntheses of chiral isotopically labelled valine. The first uses a chemical reaction of known stereochemical course in combination with a traditional resolution of a diastereoisomeric salt to incorporate the chirality (Baldwin et al., 1973), and the second introduces the chirality by means of an enzyme-catalysed step (Kluender et al., 1973). Both syntheses are conceptually quite short but require many functional-group manipulations to obtain the required product, and the steps significant for the generation of chirality are shown in Fig. 4.9. In both cases, the introduction of chirality involved carboxylic acids which were subsequently reduced to methyl groups. Baldwin's chemical route relies upon the stereospecific reduction of trans-2-methylcyclopropanecarboxylate (step g) to generate the chiral isopropyl group: the stereochemical course of ring-opening reactions is usually well defined and is a reliable strategy for the control of stereochemistry in an organic synthesis. Although this synthesis afforded a chiral isopropyl group (100% optically pure, 93% $[^{13}C]$-labelled), the formation of the other chiral centre at C-2 was not controlled and a mixture of diastereoisomers was obtained. The use of a mixture is, however, of little consequence to the subsequent biosynthetic experiment because the enzymes that synthesize penicillins will accept only the natural 2S isomer as substrate; the other isomer may bind to the enzymes, but it will not be transformed. There is still the challenge of synthesizing (2S)-valine containing a chiral isopropyl group and this challenge has been met by Sih using the enzyme β-methylaspartase. This enzyme catalyses the addition of ammonia to the

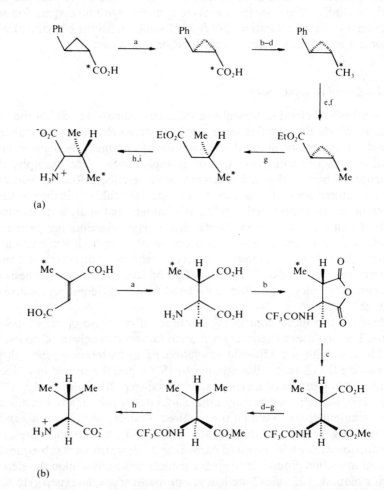

Fig. 4.9 Chemical and enzyme-based syntheses of chiral [^{13}C]-labelled valine. (a) Baldwin's synthesis. Reagents: a, resolve-quinine salt; b, LiAlH$_4$; c, MsCl; d, LiAlH$_4$; e, O$_3$; f, CH$_3$CHN$_2$; g, Li/NH$_3$; h, PCl$_3$ Br$_2$; i, NH$_3$. (b) Sih's synthesis. Reagents: a, β-methylaspartase with recycling of unreacted starting material; b, CF$_3$CO$_2$H; c, dry MeOH – the major product (80%) is shown; d, B$_2$H$_6$; e, MsCl/py; f, NaI/acetone; g, Pt/H$_2$; h, aq. HCl.

olefinic double bond of methylfumaric acid (step a). If the methyl group is [^{13}C]-labelled, then *trans*-addition of ammonia from one enantiotopic face of the substrate as controlled by the enzyme generates the *S*-configuration at both centres. Subsequent chemical modifications then led to a specimen of chiral isotopically labelled valine, 90% [^{13}C]-enriched and enantiomerically pure. Interestingly, a relative of this enzyme, aspartase, is used industrially for the synthesis of aspartic acid, which has value for enriching foodstuffs and animal feeds (Chibata, 1978). The enzyme preparation used industrially occurs intracellularly in *Escherichia coli*, and the cost of isolating it on a large scale led the Japanese scientists to immobilize whole *E. coli* cells. In this way, the cost of producing aspartic acid enzymically was reduced to 60% of that for batch fermentation.

In most syntheses of chiral compounds, a single synthetic step can be identified as the key to controlling the ultimate stereochemical outcome of the synthesis. Sometimes in a chemically based synthesis, a ring-opening or -closing reaction which can take only one course, either for stereoelectronic reasons or because of geometrical constraints, is crucial. An alternative device for controlling chirality is to use a severely hindered chiral reagent or substrate (see below). The synthetic chemist will plan his strategy around this key step. There is no alteration to this strategy if a biochemical method is used to introduce chirality; the enzyme-mediated step simply becomes the focus of the synthesis and dictates the subsequent strategy. It is not uncommon for two centres of chirality to be generated by an enzyme, as was the case with valine and the synthetic use of yeast described above. In this respect, biochemical methods can again compete with stereochemically controlled ring-opening reactions widely used in conventional syntheses. For the preparation of isotopically labelled compounds, chemical and biochemical methods are therefore valuable alternatives. Further examples have been discussed in a more extensive review of this topic by Whitlock (1976).

4.4.2   Chiral intermediate preparation

I have just alluded to the significance of steps involving the generation of chiral centres in a synthesis and the design around a key chiral intermediate. There is, of course, no need for the chirality to be introduced by isotopic labels and in most cases, labelling is not wanted. Examples in which chiral reductions are carried out by enzymes are numerous and are summarized in Table 4.1. Some general features of these reactions are noteworthy. Although as we have seen, chemical reagents can effect chiral reduction at carbonyl groups in monofunctional compounds, where polyfunctionality occurs, enzymes are often more effective. The contrast is well made if the ability of yeast to reduce α- and β-dicarbonyl compounds (Deol *et al.*, 1976) with Ohno's biomimetic chiral NADH analogues described in the previous chapter.

Table 4.1 Some generally applicable reactions catalysed by enzymes

Reaction			Enzyme	Reference
ArCDO $\longrightarrow$ ArCHDOH		S	HLADH	Jones and Beck (1976)
ArCOCO$_2$Et $\longrightarrow$ ArCH(OH)CO$_2$Me		$R(-)$	yeast	
		$S(+)$	yeast	Deol *et al.* (1976)
		cis	yeast	
			Enoate reductase	Simon *et al.* (1981)

The properties of the enzyme enoate reductase described by Simon *et al.* (1981) are striking in this context: a wide variety of structures is accepted by the enzyme and two chiral centres can be introduced simultaneously.

Enzyme-catalysed addition reactions are also very useful for introducing chirality (Table 4.2), not only into amino acids as we have already seen, but

Table 4.2 Some stereospecific addition reactions catalysed by enzymes. Those marked * have been carried out on a kg scale

Reaction	Enzyme	Reference
	Aspartase*	Chibata (1978)
	Fumarase	Hill and Teipel (1971)
	Fumarase	
ArCHO $\longrightarrow$ ArCH(OH)CN (R)	β-Hydroxynitrilolyase*	Becker *et al.* (1965)

also into hydroxy acids and notably benzaldehyde cyanohydrin. The last example has been mimicked by a chemical model (see below), but in practical synthetic terms, the enzyme is superior. Notably, this enzyme has also been used on a large scale in immobilized form and will accept a wide range of aliphatic, aromatic and heterocyclic aldehydes.

A third class of reactions in which enzymes generate chirality concerns the hydrolysis of esters and amides. In this case, the maximum yield from a racemic mixture is only 50% because the enzyme is able to hydrolyse only one enantiomer leaving the other untouched. At first sight, this appears wasteful but the recovered starting material can be racemized and recycled leading eventually to high conversions in high optical purity and the work has now been extended to β-chlorolactates and to glycidates (Hirschbein and Whitesides, 1982). In any case, the yield is at least as good as a conventional resolution via diastereoisomeric salts. In complex natural product syntheses, it is usual to find such enzyme-catalysed steps early in the reaction sequence where a loss in yield is more tolerable. Typical in this regard is the synthesis of chiral azaprostaglandins by a group of chemists at Wellcome (Caldwell *et al.*, 1979; Fig. 4.10). The deacylase used has wide specificity for α-amino acids and in this respect is valuable in synthesis. Another useful peptidase for resolving amino acids bearing aromatic α-substituents is chymotrypsin.

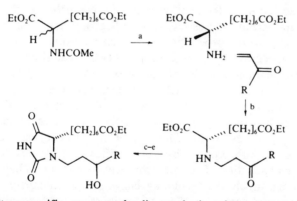

Fig. 4.10 Stereospecific azaprostaglandin synthesis using an enzyme-generated synthon. a, Hog kidney acylase; b, R = alkyl, branched alkyl, cycloalkyl; c, NaBH$_4$; d, HNCO; e, H$_3$O$^+$.

In the reactions just described, the chirality was introduced into the acidic component of the ester or amide but so far relatively little success has been obtained in the enzymic resolution of esters of chiral alcohols. An isolated case has been described in an ester of a derivative of a substituted 2-indanol (Kawai *et al.*, 1981). Occasionally, the most improbable substrates have been resolved. For instance Reid *et al.* (1967) showed that a racemic sulphite ester

Table 4.3 Some stereospecific enzyme-catalysed hydrolyses

Reaction	Enzyme	Reference
EtO₂C–CH(R)(H)–NHCOCH₃ → EtO₂C–CH(R)(H)–NH₂	Hog kidney deacylase	Caldwell *et al.* (1979)
RCHCO₂Me(NHCOCH₃) → RCHCO₂H(NHCOCH₃)	Chymotrypsin	Jones and Beck (1976)
(bicyclic CO₂Me, X, =O structure) → (CO₂Me, H, X, =O)	X = CH₂, NH, O Chymotrypsin	Jones and Beck (1976)
AcO,H,Br indanyl → HO,H,Br indanyl	*Rhizopus nigricans*	Kawai *et al.* (1981)
MeO₂C···CO₂Me → HO₂C···CO₂Me and MeO₂C···CO₂H	Pig liver / *Gliocladium roseum*	Chen *et al.* (1981)

could be resolved using the peptidase, pepsin. Nevertheless, the most usual substrates are carboxylate esters, and Sih's group has extended the techniques available to the production of two chiral centres in one incubation (Table 4.3; Chen *et al.*, 1981). The half-esters were readily converted into the enantiomeric lactones in high optical purity.

The use of acylases in the industrial production of L-amino acids is one of the classics of industrial enzyme chemistry. As long ago as 1953, Japanese scientists were hydrolysing acyl DL-amino acids with an aminoacylase from the mould *Aspergillus oryzae* (Chibata, 1978). This enzyme combines high catalytic activity with broad substrate specificity, properties that are ideal for synthetic application. The early work was carried out in batch processes but the labour involved in isolating enzymes and in purifying products severely hampered operations. This difficulty stimulated the initial development of immobilized enzymes. Not only were chemical factors such as enzyme stability and efficiency investigated but also the new engineering problems of handling enzymes on a large scale were tackled. One of the main streams of modern biotechnology thus has its source in this work. As with the immobilized *E.coli* cells mentioned earlier, the saving in operating costs using immobilized enzymes was more than 40%. Industrial processes also use enzymes to hydrolyse amides in more complex structures; the selective cleavage of the side chain in penicillins leaving the important β-lactam intact is perhaps the best-known example.

Apart from some of the reduction reactions, there are currently no good chemical competitors to these biochemical methods for the synthesis of chiral intermediates. Chemists have expended a great deal of ingenuity to make use of readily available chiral natural products (amino acids, sugars, terpenes, etc.) as intermediates in syntheses or to provide chiral templates for asymmetric transformations (Szabo and Lee, 1975). Clearly by the use of biochemical reagents, the range of chiral building blocks for organic synthesis can be greatly extended. To conclude this section, an example of a natural product synthesis from Jones' group which applies HLADH chemistry will be outlined. Several syntheses of the insect pheromone, grandisol, have been described but in terms of yield and optical purity of the product, this synthesis is the best that has yet been achieved and its success depends entirely upon the early HLADH oxidation to form the chiral lactone (*4.1*, Fig. 4.11). Once this compound was in hand, it was a matter of careful elaboration and transformation of intermediates to obtain grandisol (Jones *et al.*, 1982). A further eloquent example of the usefulness of enzymes to generate chiral synthons is a synthesis of (+)-4-twistanone from enzymically reduced decalones (Dodds and Jones, 1982).

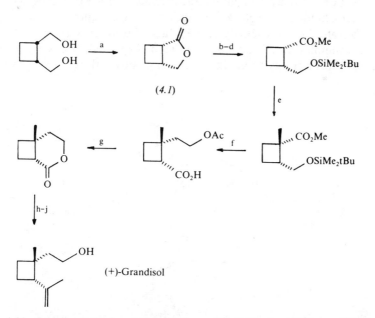

Fig. 4.11 Jones' synthesis of grandisol. a, HLADH; b, KOH; c, CH$_2$N$_2$, d, tBuMe$_2$SiCl; e, LDA, MeLi; f, several steps extending upper side chain via Wittig reaction and hydroboration and modifying lower substituent by deprotection and oxidation; g, 5% H$_2$SO$_4$; h, MeLi; i, Ac$_2$O; j, LiAlH$_4$.

4.5 LATE-STAGE FUNCTIONAL-GROUP MODIFICATION

If the overall success of a synthesis depends crucially upon the strategy, especially where chirality is involved, then the bringing of an elegant strategy to fruition relies upon the chemist's ability to modify selectively functional groups in the molecule. At this late stage in a synthesis, difficulties often centre around the introduction of isolated chiral centres remote from control elements such as rings; differentiation of similarly reacting groups should have been controlled by the use of suitable protecting groups. However, there is no reason why enzymes cannot also be useful in this phase of a synthesis and there are cases where they have been more effective than chiral chemical reagents. For example, in a prostaglandin synthesis, Sih *et al.* (1975) (Fig. 4.12) found that it was impossible to reduce the cyclopentatrione selectively with a chiral rhodium phosphine complex that had proved successful elsewhere. His experience with micro-organisms led him to expect that biochemical methods would be successful and, by choice of a suitable micro-organism, it was possible to produce either enantiomer at will.

The reader will by now be familiar with the notion that enzymes control selectivity by bringing the reactants together in the required juxtaposition. Perhaps the most successful but remote chemical relative of this device is the use of steric hindrance to block the attack of a reagent upon a particular part of a substrate. The unreactivity of the β-face of steroids was one of the first general examples of this phenomenon to be exploited, but chemists have developed this strategy to attain chiral reductions with selectivity rivalling

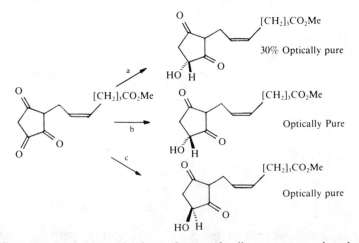

Fig. 4.12 Enantioselective reductions of prostaglandin precursors using chemical and microbiological methods. a, Chiral rhodium complex; b, *Dipodascus uninucleatus*; c, *Mucor rammanianus*.

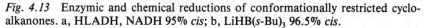

Fig. 4.13 Enzymic and chemical reductions of conformationally restricted cyclo-alkanones. a, HLADH, NADH 95% *cis*; b, LiHB(*s*-Bu)₃ 96.5% *cis*.

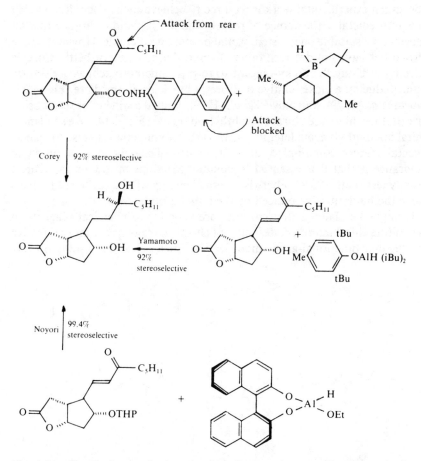

Fig. 4.14 Chemical strategies in reduction of enones in prostaglandin synthesis.

enzymes (Krishnamurthy, 1974). Simple substrates such as cycloalkanes can be reduced with high selectivity by complex borohydrides such as lithium or potassium tri-*s*-butyl borohydride; clearly attack by such a bulky reagent will take place more readily from the less-hindered side of a molecule to give, for instance, the axial alcohol (Fig. 4.13). As an alternative to letting a bulky reagent generate selectivity, it is possible to insert a large group into the substrate that will temporarily block attack from one direction. Both approaches have been exploited in controlling the stereochemical course of reduction at C-15 in the prostaglandin system (Fig. 4.14). In Corey's method (Corey *et al.*, 1972) the side-chain ketone was reduced by a chiral borane with high selectivity from the rear to yield the S-isomer (92% optically pure). Anyone who takes the trouble to read Corey's paper will appreciate not only how much experimental work is required to achieve such selectivity but also how unpredictable the choice of reagent can be. It could be argued that an enzyme, if a suitable one existed, would be more predictable. However, these difficulties have not deterred others from trying to equal or better Corey's selectivity. Using the bulky-reagent strategy, Yamamoto achieved a similar result reducing a close relative of Corey's blocked intermediate (Fig. 4.14; Iguchi *et al.*, 1979), and Noyori *et al.* (1979) claimed a still better result using an extremely hindered chiral binaphthyl aluminate (Fig. 4.14). We shall meet chiral binaphthyls again later in discussion of biomimetic systems. An experimental feature common to all of these strategies depending upon steric hindrance is that it is essential to conduct reactions at low temperatures, usually less than −78°C otherwise thermal energy will be sufficient to overcome the hindrance introduced by the bulky group.

It might be claimed that enzymes are superior to chemical reagents in generating chiral intermediates but that chemical reagents are more effective in late-stage functional-group modification. Such a clear-cut distinction may

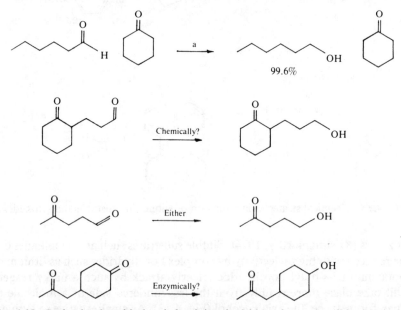

Fig. 4.15 Reduction by a hindered aluminium hydride and some possible chemical/ enzymic selective reactions. a, Li(Et₃CO)₃AlH, −78°C, 3h.

have some limited value as a crude generalization, but things are rarely so simple. The cold non-polar conditions required for the use of chiral boranes and hindered compounds will not suit all substrates. Although derivatization can often overcome solubility problems, it is best to treat each case on its merits. After all, there are examples of enzyme-catalysed reactions for which chemical reagents have no competitor, for instance Simon's enoate reductase. Equally there are cases where chemical selectivity has the advantage. An intriguing recent example suggests many possible interplays of chemical and enzymic reduction. Krishnamurthy (1981) has shown that the very hindered aluminate, lithium tris-3-ethylpentyloxy aluminium hydride is too bulky to reduce even unhindered cycloalkanones at low temperature, but will reduce aldehydes (Fig. 4.15). Bearing in mind the current understanding of HLADH selectivity, it is interesting to speculate upon how the other selective transformations of Fig. 4.15 might be accomplished.

4.6 BIOMIMETIC CHEMISTRY IN SYNTHESIS

A study may be termed 'biomimetic' when the chemist consciously designs his programme based upon a feature of the chemistry of a natural system. The system could be a transport mechanism, a biosynthetic pathway, a group of receptors, a nucleic acid as well as an enzyme. For the purposes of this chapter, we are concerned with synthetically applicable biomimetic chemistry. Much has already been written concerning the modelling of synthetic strategy upon biosynthetic pathways, the so-called biogenetic type synthesis (Scott, 1976; Suckling et al., 1978), but the significant comparison for this discussion is between enzymes used in single synthetic steps and biomimetic reactions that attempt in some way to profit from our understanding of enzyme mechanisms. Serious attempts by chemists to gain selectivity by means of enzyme mimicking is a relatively recent development despite Emil Fischer's optimism (see Chapter 1) and a great deal of credit for its development goes to Ronald Breslow. His two reviews of his work (Breslow, 1972 and 1980) are major signposts in the field.

There are several approaches to biomimetic systems. In many cases the objective has been to match the catalytic efficiency of enzymes. Although such work has had a major influence upon the subject by highlighting important concepts, the emphasis upon hydrolysis reactions of p-nitrophenyl esters has minimized the synthetic value. Exceptions to this limitation may be noted in those cases where stereoselective reactions have been attempted. Indeed the essence of synthetic biomimetic chemistry is the control of selectivity by some form of binding of the substrate to the enzyme mimic. As in the preceding sections in this chapter, comparisons will be made where appropriate with closely related enzyme-catalysed reactions. However, in biomimetic chemistry, a new degree of flexibility is introduced.

Although the reaction is based upon enzyme chemistry, there need not be a known enzyme that catalyses the reaction in question. Hence a more meaningful comparison will be between the biomimetic system and a conventionally designed chemical reaction.

Several key systems capable of providing useful binding for biomimetic synthesis have been discovered. The first to be investigated extensively was the cyclodextrin system. *Cyclodextrins* are cyclic oligomers of glucose, β-1,4-linked, containing six, seven or eight glucose molecules. They possess a toroidal structure with an outer coat of hydroxyl groups and a largely non-polar inner cavity (Fig. 4.16). It is the inclusion of non-polar substrates within this cavity that gives cyclodextrins their useful biomimetic properties. Secondly, *crown ethers* have developed a substantial chemistry. The first of these molecules to be described were macrocyclic derivatives of ethylene oxide (Fig. 4.16). Unlike cyclodextrins, which bind non-polar substrates, crown ethers complex with ions; originally they were significant for their ability to bind metal cations but the discovery that primary ammonium cations will also bind has opened up an organic chemistry. More recently, analogues have been described that will bind anions. The third major class of biomimetic systems is less well defined and concerns *micelles*. The structure of these multimolecular aggregates of detergent-like molecules is still a matter of debate but they have demonstrably interesting biomimetic properties as we shall see. In addition to these major groups several other systems have been studied, notably small peptides.

The essence of control of selectivity by any of the systems outlined above is ordered binding of the substrate. All types of chemical forces can be used to control binding. In some of the first successful biomimetic syntheses, covalent bonding was used to attach a reagent to a substrate, in this case a

Cyclodextrin   Crown ether

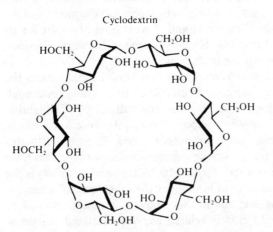
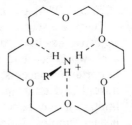

Fig. 4.16   Some building blocks for biomimetic systems.

steroid, in a specific manner. Crown ethers rely essentially upon ionic or dative covalent bonding. However, weaker chemical forces can also be significant; cyclodextrins use predominantly hydrophobic binding and micellar systems mix ionic and non-polar interactions with hydrogen bonds. In other words, all the forces available to enzymes are used in some form by biomimetic chemistry.

As I have said, biomimetic synthesis is a relatively young field and many of the examples to be described are pioneering. From the point of view of laboratory-scale synthesis, conventional chemical techniques and enzymic procedures will often have the advantage but if simple and effective selective biomimetic systems can be devised, they will be very significant on an industrial scale and will compete with immobilized enzymes and genetically engineered preparations. It may seem that the target for selectivity in biomimetic systems is unattainably high at first sight. However, as was first pointed out by Guthrie (1976), even a modest increase in yield may be important commercially. Consider the data in Table 4.4 which illustrates the simple case of a substance S being transformed into a product P and unwanted by-products. Taking an arbitrary 90% yield as acceptable, Table 4.4 clearly shows that improvements in selectivity of as little as 10- to 1000-fold may well be useful. Since the catalytic effects of enzymes are commonly estimated as being between 10^5 and 10^9, to harness but a small fraction of their selectivity in a biomimetic system could be valuable. With this encouragement, we can proceed to discuss synthetic biomimetic chemistry.

Table 4.4   A substance S is transformed into a product P and unwanted by-products:

$$S \xrightarrow{k_1} P \qquad S \xrightarrow{k_2} \text{Waste}$$

k_2/k_1	Yield without catalysis, %	Catalysis required for 95% yield	Typical reaction
1	50	9	Aromatic substitution
9	10	81	
99	1	891	Aliphatic radical substitution
199	0.5	1791	
0.25	80	2	
0.11	90	11 (for 99% yield)	

4.6.1   Covalent control of selectivity

The use of ring systems to control stereoselectivity in synthesis has been a major strategy (Norman, 1982). Six-membered rings, whose conformations lead to well-understood stereoelectronic effects have been especially

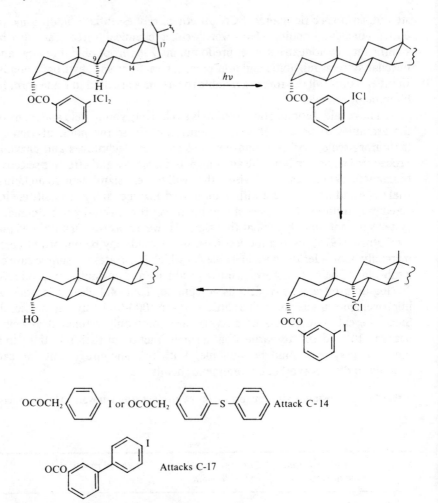

Fig. 4.17 Breslow's directed chlorination in steroid synthesis.

important as have the constraints geometrically imposed by the opening of small rings. However, little attention had been paid to the possibility of using large rings in the control of regioselectivity in synthesis until biomimetic studies of steroid functionalization were begun by Breslow. It had been known for many years that micro-organisms were capable of carrying out specific oxidations at chemically non-activated positions in the steroid-ring system. Clearly control of selectivity is provided by specific binding of the substrate to the enzyme's active site and Breslow's target was to construct a simple chemical system that would locate the reagent specifically with respect to the rigid steroid-ring system (Fig. 4.17). The most susceptible saturated carbon atoms for attack are the tertiary centres at C-9, -14 and -17. The hydrogen atoms at these centres are all α and axial and can therefore be abstracted by a radical approaching from the α-face. The problem then resolves itself into positioning the reagent appropriately. Experiments in which substituted benzophenones were attached via ester linkages to a 3-hydroxyl function and hydrogen abstraction was initiated photochemically showed that the strategy was feasible. The technique was then developed into a synthetic chlorination process by means of substituted iodobenzenes or diphenylsulphides which place a chlorine atom close to either of the target hydrogen atoms. The reaction sequence is initiated through hydrogen abstraction by the strategically placed chlorine atom which removes the nearby hydrogen atom. Pairing of the tertiary radical with another chlorine atom then leads to the selectively substituted products. Provided that dilute solutions were used to minimize the chance of intermolecular reactions, high selectivity was obtained. Indeed if the pendant chain bearing the chlorine atom was longer than the steroid rings, no reaction occurred. The chlorinated steroids could then be further transformed to alkenes and, in the case of the 17-chloro compound, it was possible to degrade the side chain leading to valuable precursors for the synthesis of steroid hormones. Recent developments of these techniques have resulted in a catalytic multiple chlorination of steroids (Breslow and Heyer, 1982).

To make a direct comparison with enzyme-catalysed reactions, it would be nice to be able to insert oxygen atoms as well as chlorine. Although hydroxylation has not yet been realized, largely because of the difficulty of designing a suitable reagent, epoxidation can be controlled selectively over remarkably long distances. Sharpless and Verhoeven (1979) showed that t-butyl hydroperoxide was a very useful epoxidizing agent in the presence of transition-metal complexes of vanadium and molybdenum. In these reactions, the metal acted as a template and co-ordinated with both the hydroperoxide and a hydroxyl group in the substrate. For instance, the double bond of farnesol or geraniol closest to the alcohol could be specifically epoxidized (Fig. 4.18). Breslow adapted his template to include a tertiary alcohol to which a metal could bond. It was then possible to direct epoxidation to a remote position in the steroid substrate.

Steroids, with their rigid ring structure, are ideal substrates for this kind of biomimetic strategy. The predictability of the behaviour of the template systems is also an advantage. However, the application of similar techniques to more flexible molecules has invariably resulted in the production of mixtures. The problem of selective functionalization of simple alkanes, a problem of industrial significance in the production of primary alcohols for detergents and plasticizers, has still to be solved by biomimetic chemistry, although selective terminal hydroxylation of alkanes can be accomplished by

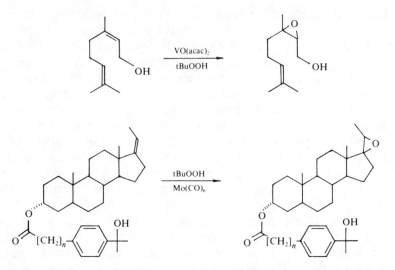

Fig. 4.18 Template-directed epoxidations: $n = 1$ epoxidation occurred, $n = 2$ no reaction.

micro-organisms. As has already been mentioned, hydroxylation reactions of steroids have been used in industrial syntheses of corticosteroids for many years, but despite considerable interest, the microbiological hydroxylation of simple aliphatic compounds has not been successful so far on a commercial scale.

4.6.2 Cyclodextrins

The major area of application of the binding properties of cyclodextrins has been the development of systems aimed at mimicking the high rate enhancements of enzymes. In this respect, research has been spectacularly successful (Breslow, 1980; Tabushi, 1982); however, synthetic studies have been less extensive. The cavity in the cyclodextrin torus is a suitable size to bind mono- and bi-cyclic aromatic rings. For example, anisole and other benzene derivatives with one small substituent, bind so that the substituent is buried in the cavity. The *para* position is usually the most exposed in the cyclodextrin complex and can be attacked by a suitable reagent (Breslow, 1972, 1980). Thus when the anisole complex was reacted with a solution of hypochlorous acid, the peripheral hydroxyl groups of the cyclodextrin exchanged with the hypochlorite and formed a chlorinating agent ideally placed to cause *para* substitution with high selectivity. Interestingly, the most closely related enzyme-catalysed reaction to this chlorination occurs with inferior selectivity close to the ratio obtained with hypochlorous acid in the absence of the cyclodextrin.

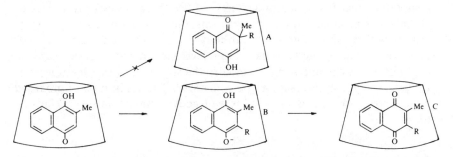

Fig. 4.19 Multiple catalytic effects of a cyclodextrin on naphthalene diol oxidation. $R = CH_2CH=CH_2$, CH_2, $CH=CHCH_3$ or $CH_2CH=CMe_2$. A, Alkylation to form this complex is too hindered. B, Cyclodextrin increases anion concentration and catalyses alkylation. C, Oxidation of diol occurs by O_2 affording H_2O_2: further oxidation of quinone by H_2O_2 is prevented by cyclodextrin.

A rare case of selective substitution by means of cyclodextrin complexation was described by Tabushi *et al.* (1979); the nearest enzyme-catalysed relative to the reaction studied is the prenylation of phenolic natural products, but the enzymes that catalyse this reaction do not appear to have been isolated. Tabushi was interested in a synthesis of vitamin K and he required to catalyse and control the alkylation of a phenol with respect to O- or C-alkylation. His solution was to alkylate the dinaphthol (Fig. 4.19) complexed with a cyclodextrin; concomitant oxidation led to the required intermediate naphthoquinone. A kinetic analysis showed that alkylation was catalysed by the cyclodextrin and that complexation also prevented further oxidation of the product. Selectivity in the alkylation of naphthols can also be controlled by micellar systems as will be described below.

Stereoselectivity is also possible using cyclodextrin derivates although the origin of the selectivity is unclear. Breslow found that pyridoxal attached to β-cyclodextrin was effective in catalysing transamination of aromatic α-keto acids to the corresponding α-amino acid (Breslow *et al.*, 1980). Preferential reactivity towards keto acids with aromatic groups that can bind in the cyclodextrin cavity was observed and in such cases, the enantiomeric selectivity was also high. A potentially very significant synthetic application of cyclodextrin chemistry concerns the Diels–Alder reaction (Rideout and Breslow, 1980). In reactions where both reactants can complex together within the cyclodextrin's cavity, substantial catalysis occurred and the possibility clearly exists of controlling regioselectivity in such complexed cycloadditions. Similar rate enhancements were also observed in solutions of ionic detergents, an interesting observation in the context of micellar systems described below.

These few examples prove the synthetic utility of cyclodextrins. It might seem that the cyclodextrin system is somewhat limited in the range of compounds that it can accept as guests. Since there are three ring sizes of cyclodextrin produced by micro-organisms, the diameter of the cavity can be chosen within reasonable limits, although it cannot yet be easily varied. However, synthetic methods have been developed to extend or decrease the depth of the cavity and hence to modify the substrate specificity at least as far as ester hydrolysis reactions are concerned. The application of modified cyclodextrins to synthesis is a subject of active research in several laboratories.

4.6.3   Crown ethers

The ability of the first crown ethers to bind inorganic ions was recognized as long ago as 1967 and it was soon shown that the strongest binding was obtained for those cations whose ionic radii most closely matched the available space between the donor oxygen atoms. Potassium bound especially strongly to ethers containing six donor oxygen atoms spaced equally between dimethylene bridges. Since the ammonium cation has an ionic radius very similar to potassium, it was no surprise to find that stable ammonium complexes could be prepared. It was then but a small step to show that primary alkyl ammonium cations would also bind, and from this result, the organic chemistry of crown ethers has grown. A direct synthetic application of the ability of crown ethers to bind primary ammonium cations was demonstrated by Barrett et al. (1980). Although a primary ammonium cation binds, largely with the aid of hydrogen-bonding of the three hydrogen atoms, with three of the donor oxygen atoms of the ligand, secondary ammonium cations are both too bulky and too ineffective in hydrogen-bonding to complex with the ligand. It was therefore possible to protect a primary ammonium cation selectively in the presence of a secondary cation with a crown ether. Unsymmetrical acylation was then carried out in vastly improved yield over a comparable uncontrolled reaction (Fig. 4.20).

A major area of interest in crown ether chemistry has been the introduction of chirality into the macrocyclic ether. Diols from carbohydrate

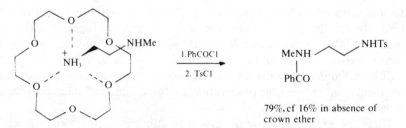

Fig. 4.20   Crown ether-controlled acylation of a diamine.

derivatives can be converted into crown ethers, Cram's use of chiral binaphthyl diols (Fig. 4.21; Cram, 1976) has been especially effective. The chirality
in these compounds, like biphenyls, is due to restricted rotation about the
central C–C bond. The first application of these ligands was in the formation of enantioselective complexes with derivatives of amino acids. By
means of a U-tube transport apparatus, it was possible to effect a resolution
of the racemic amino acid analogous to the use of the specific esterase
enzymes described earlier. Recently, the potential of these chiral crown
ethers has been greatly extended by the demonstration of asymmetric
Michael addition reactions to methyl vinyl ketone (Cram and Sogah, 1981).
Although similar reactions have been carried out before, using chiral
alkaloids to provide the asymmetric induction, the best examples of crown
ether-mediated reactions proceed with almost complete stereoselectivity. It is
essential to use the potassium enolate as the nucleophile so that a
host–K$^+$–substrate$^-$ complex can be formed. Since potassium ions bind
strongly to the crown ether, the enolate can be maintained in a chiral
environment and will add to methyl vinyl ketone to give a product in a high
state of optical purity. As with the resolution experiments, the asymmetric
induction relies upon the attainment of the crown ether complex in which
steric interactions between bulky groups on the host and guest molecules are
minimized. It was also extremely important to operate at low temperature
and to choose reaction conditions carefully so that the differences in energy
between the possible diastereoisomeric complexes could have most effect.
Thus in the reaction illustrated in Fig. 4.21, at −78°C, the product was

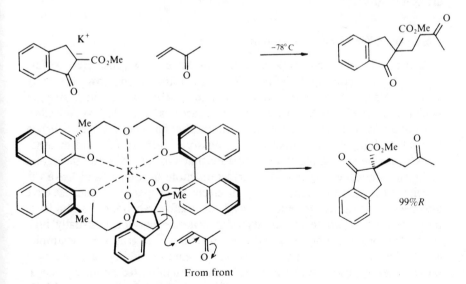

From front

Fig. 4.21 Enantioselective Michael addition mediated by a crown ether.

obtained in 99% enantiomeric excess whereas at 25°C, the optical purity was reduced to 67%.

Crown ethers, unlike cyclodextrins, have the advantage of being man-made molecules whose structures can be designed to suit the problem at hand. Although not directly relevant to synthesis, crown ethers capable of mimicking enzymic enantioselective hydrolysis and reduction reactions have been synthesized. Some ingenious molecules have been designed incorporating external control of complexing ability; photosensitive groups permit the switching on and off of complexation in some compounds and, in others, a second metal ion can control the conformation of the crown ether again controlling complexation. The latter example can be compared to allosteric control of the activity of an enzyme. In addition to the crown ethers that bind main-group cations, derivatives containing β-diketones have been prepared to interact with transition metal ions. Further variants bind phosphates through macrocyclic amidines. Such compounds mimic a range of biological receptors and can be constructed to accommodate bifunctional compounds with structures similar to neurotransmitters (Sutherland, 1982). Molecules that are capable of binding two functional groups might clearly bind two substrates. In this way, a class of artificial enzymes might be designed. Recent work has shown that it is possible to synthesize functional analogues of ionophores, the components of cell membranes responsible for the transport of inorganic ions such as potassium and chloride. Thus potential of this family of compounds is much wider than the synthetic field and applications to switching devices and to selective artificial transport systems are conceivable.

4.6.4   Micellar and related systems

In contrast to the biomimetic systems that have been considered so far, micelles are structurally ill-defined. It has, of course, been known for many years that surfactants with greater than ten carbon atoms in a straight aliphatic chain will readily agglomerate in aqueous solution to form micelles. The gross physical chemical properties of micelles have been well established by such techniques as light-scattering measurements. In this and in other ways it has been possible to demonstrate that typically 50–100 molecules make up a micelle and that the surfactant molecules enter and leave the micelle at rates greater than $10 \, s^{-1}$. The electrical properties of micelles have also been well studied; micellar catalysis of organic reactions has to a large part been attributed to the ability of micelles to develop unusually high concentrations of counterions in the region of the head group. For example, a hydrolysis reaction mediated by hydroxide anion would be catalysed by a micelle bearing positively charged head groups but inhibited by one with negatively charged heads (Bunton, 1976). Despite careful studies with skilfully

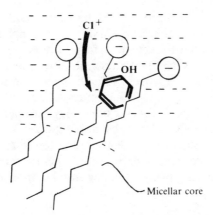

Fig. 4.22   Orientation of phenol in micellar solution and its effect upon chlorination.

selected probes, the detailed structure of micelles is uncertain, which is a major disadvantage to the design of synthetic systems incorporating micelles as the binding group. One of the most useful representations is due to Menger (1979) who suggested on the basis of spectroscopic and model-building studies that micelles are irregular spheroids in which the component surfactant molecules enter and leave through grooves in the surface. Water penetrates up to seven carbon atoms into these grooves, and from the synthetic point of view, it is attractive to imagine the grooves as convenient binding loci for substrates.

With these difficulties in mind, it is not surprising that studies of the application of micellar systems to biomimetic synthesis have proceeded empirically (Suckling, 1981). Micellar systems have the virtue that they are simple to construct but the price for this facility is the lack of a clear structure, like a cyclodextrin or crown ether, on which to design a binding and an active site. The most encouraging suggestions that micelles might be useful in synthesis came from several laboratories where it was shown that in aliphatic chemistry, spectacular control of reductive mercuration of terpenoid molecules could be achieved (Link *et al.*, 1980), and orientation in electrophilic and radical aromatic substitution can be influenced by micelles. We were able to demonstrate by means of high-field nuclear magnetic resonance (NMR) spectroscopy that micelles cause small polar aromatic substrates to be oriented as shown in Fig. 4.22. If the reagent approaches the substrate from the polar aqueous phase, then it is easy to see how an increase in *ortho* substitution can result. In a number of cases, the chemical selectivity corresponded well to the orientation of the substrate suggested by the NMR spectra. These observations begin to provide the basis for the design of more effective selective substitution systems.

There are two obvious approaches both of which add a degree of control to the semi-ordered micellar system. Firstly, several molecules of surfactant can be covalently bonded together to form a molecule bearing tentacles. Menger *et al.* (1981) was the first to describe such a molecule (Fig. 4.23); his six-armed 'hexapus' was capable of solubilizing a molecule as large as cholesterol in water. Our own tentacle molecules so far contain three arms and can be shown to bind small aromatic molecules like phenol both by inhibiting chlorination reactions and by substantial changes in the NMR spectrum analogous to those found in micellar solution (Suckling, 1982; Fig. 4.23). Both of these molecules are, however, prototypes for selective functionalization systems in that they possess binding sites but have still to be equipped with active sites.

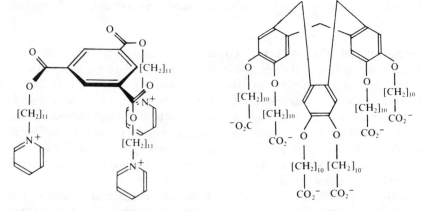

Fig. 4.23 Novel tentacle molecules that may become the basis of biomimetic systems.

The second way to improve the selectivity of micellar systems is to localize the reagent at a defined position within the micelle. The clearest indication of the importance of this comes from our work on the chlorination of phenol in micellar solution. As Fig. 4.22 illustrated, we were able to show that phenol is solubilized on average close to the head group such that the polar hydroxyl group points into the aqueous phase. Clearly to obtain selective *ortho* substitution, the reagent must be inserted at about C-3 of the micelle. This was achieved using a functionalized stearic acid bearing a tertiary alcohol (Fig. 4.24; Onyiriuka and Suckling, 1982); the hypochlorite from this alcohol caused chlorination of phenol to occur with close to complete selectivity for *ortho* substitution. The most important conclusion from this work is that it is possible to obtain high selectivity even with a relatively poorly defined binding site provided that the reagent is properly positioned with respect to the substrate. When compared with conventional reagents, these micellar

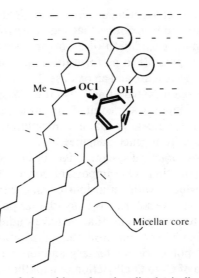

Fig. 4.24 Chlorination of phenol by a functionalized micelle.

systems have the major potential of being able to carry out a selective transformation in principle in one step. It is always possible to devise a lengthy
synthesis of a substituted aromatic compound, and systems developed from
the results outlined above should offer an elegant and predictable alternative.

A major problem outstanding in this work is the design of a specifically
inserted hydroxylating agent. Several hydroxylating enzymes employing
transition metal complexes, often haems, as generators of hydroxyl reagents
are well known, but these oxidizing agents are too large and complex to be
incorporated easily into micelles and their relatives. Nevertheless, hydroxyl-
ases themselves can be used effectively in synthesis as some recent studies on
the production of tyrosine from phenylalanine show (Klibanov *et al.*, 1981).

Digressing briefly on the subject of haem-containing enzymes, it is worth
noting the use of the enzyme chloroperoxidase in selective synthesis. In the
presence of hydrogen peroxide and chloride, this enzyme catalyses the
insertion of chlorine as an electrophile into an organic substrate, usually an
alkene or an electron-rich aromatic compound. A telling recent application
concerns the large-scale synthesis of ethylene and propylene oxides. Conventional heterogeneous catalysed oxidation of the alkenes always leads to
some losses due to over-oxidation to carbon dioxide. Chloroperoxidase, in
contrast, will convert these substrates into the corresponding epoxide or
halohydrin without over-oxidation. When the epoxide is the product,
chloride is regenerated and thus is also catalytic (Neidlemann *et al.*, 1981).
The fuel for the reaction is hydrogen peroxide and, during development

work, it was felt that conventional methods of hydrogen peroxide production might be bettered by oxidases that generate hydrogen peroxide from oxygen with concomitant oxidation of an organic substrate. Research has since shown that it is possible to perform this important epoxidation using two enzymes, chloroperoxidase and an oxidase. It is significant that immobilized enzymes were again favoured and also that considerable biochemical and microbiological input was required to produce suitable strains of organisms as enzyme sources. A major contribution to this area can be expected in the future from genetic engineering.

It is easy to see how practical systems might be developed from the observations in micellar solution. One important consideration is to facilitate product recovery which, with emulsion-forming surfactants around, is troublesome. An attractive solution is to synthesize polymeric analogues of micelles. Although to date, systems that we have studied have not been able to reproduce micellar effects on aromatic substitution reactions, Brown (1980) showed that polymeric quaternary ammonium salts were useful in controlling the ratio of C- to O-alkylation in naphthols. He showed that the microenvironmental effect of the micellar analogue was to enhance the nucleophilicity of the oxygen atom and thereby to increase the proportion of O-alkylation. This result was complementary to Tabushi's observation using a cyclodextrin.

4.6.5   Other biometic possibilities

In addition to the three main groups of binding species described above, other molecules have been found to have biomimetic binding properties which may in time find a use in synthesis. Some experiments with small peptides have shown that it is possible to mimic the stereospecific enzyme-catalysed addition of cyanide to benzaldehyde using a cyclic phenylalanyl-histidine (Oku and Inoue, 1981). Many chemists have attempted to achieve stereoselective epoxidation using biomimetic chiral catalysts or simpler chiral-phase transfer catalysts. Usually base catalysed reaction of $\alpha\beta$-unsaturated ketones have been studied, and in many cases, high chemical yields and almost 100% enantiomeric excesses were obtained (Julia et al., 1980; Wynberg and Marsmann, 1980). These reactions are related to Cram's asymmetric Michael additions mediated by crown ethers.

Recently, some interesting hybrid systems have been described. Kaiser (Slama et al., 1981) has incorporated a foreign coenzyme, a flavin, into an enzyme, papain, with a well-defined binding site. This is an attractive pairing because papain is known to contain an active-site thiol group, and many natural flavoenzymes use a thioether to bind the flavocoenzyme to the active site. Although the work is in its infancy, he has been able to demonstrate high catalytic activities of the flavin in oxidation reactions in its unfamiliar environment. Alcohol dehydrogenase is, as we have seen, extremely well studied from the point of view of substrate specificity. Curiously, it also contains the charge relay system of the serine proteinases and it is therefore intriguing to wonder whether it might be persuaded to act as an esterase with suitable substrates. Zeolites are silicate minerals that possess channels or pores with a defined diameter. Also zeolites can be manufactured to contain acidic or basic surface catalytic sites. They have been widely studied for their potential in controlling large-scale rearrangement and substitution reactions of small hydrocarbons. The analogy between the pores of the zeolite and the binding site of an enzyme is obvious. Recently the use of a zeolite to catalyse the hydrogenation of fructose to mannitol has been combined with immobilized glucose isomerase to produce a one-pot conversion of glucose, which is readily available from corn syrup, into mannitol, which currently commands high prices for pharmaceutical and medical applications (Ruddleston and Stewart, 1981).

4.7 CONCLUSIONS

Writers on synthesis in recent years have stressed such qualities as convenience, predictability and reliability in their evaluation of reagents. I have tried to show how far enzymes and other biochemical systems match up to these requirements. In the case of alcohol dehydrogenase, a positive response can certainly be given to each criterion. Although few other enzymes have been so well assessed, the type of selective transformation that enzymes are good at is usually crucial to the success of a synthesis and accordingly it must be worth while to consider seriously whether a biochemical method might be suitable. There is no reason why today enzymes should not be thought of on equal terms to chemical reagents for selective transformations. Indeed if a good analogy can be found, a microbiological method might be valuable too. What about biomimetic methods? At present, synthetically useful examples are to be found in only a few cases. However, this was the case with enzymes themselves ten years ago and with the principles of design of biomimetic systems becoming daily more firmly established, there can be little doubt that they will break into synthetic chemistry within the next decade. It may be that their best field of operation will be in industrial chemistry where it is often necessary to carry out relatively few reactions supremely well. Some readers may have found the hybrid systems mentioned in the last few paragraphs esoteric but there are real opportunities in this area for imaginative work. A fusion of ideas from coenzyme chemistry (Chapter 3) and structural protein biochemistry (Chapter 8) is required. A suitable protein could provide a well-defined chiral environment within which a non-natural selective synthetic reaction can take place. The simple catalytic advantages of environment control have already been highlighted by Professor Shinkai (Chapter 3) and by Professor Breslow's cyclodextrin-mediated Diels–Alder reaction (this chapter): high performance can thus be expected from hybrid systems.

REFERENCES

Baldwin, J. E., Loliger, J., Rastetter, W., Neuss, N., Huckstep, L. L. and de la Higuera, N. (1973) *J. Am. Chem. Soc.*, **95**, 3796.
Barrett, A. G. M., Lana, J. C. A. and Tograie, S. (1980) *J. Chem. Soc., Chem. Commun.*, 300.
Battersby, A. R. and Staunton, J. (1974) *Tetrahedron*, **30**, 1707.
Battersby, A. R., Staunton, J. and Wiltshire, H. R. (1975) *J. Chem. Soc., Perkin Trans.*, **1**, 1156.
Becker, W., Freund, H. and Pfeil, E. (1965) *Angew. Chem.*, **77**, 1139.
Breslow, R. (1972) *Chem. Soc. Rev.*, **1**, 553.
Breslow, R. (1980) *Acc. Chem. Res.*, **13**, 170.
Breslow, R. and Heyer, D. (1982) *J. Am. Chem. Soc.*, **104**, 2045.
Breslow, R., Hammond, M. and Lauer, M. (1980) *J. Am. Chem. Soc.*, **102**, 421.
Brown, J. M. (1980) in *Enzymic and Non-enzymic Catalysis* (eds P. Dunnil, A. Wiseman and N. Blakebrough), Ellis and Horwood, Chichester, p. 120.
Bunton, C. A. (1976) in *Applications of Biochemical Systems in Organic Chemistry, Techniques of Organic Chemistry Series* (eds J. B. Jones, C. J. Sih and D. Perlman), Wiley Interscience, New York, Vol. 2, p. 731.
Caldwell, A. G., Harris, C. J., Stepney, R. and Whittaker, N. (1979) *J. Chem. Soc., Chem. Commun.*, 561.
Chen, C-S., Fujimoto, Y. and Sih, C. J. (1981) *J. Am. Chem. Soc.*, **103**, 3580.
Chibata, I. (ed.) (1978) *Immobilised Enzymes*, Halsted Press for Kodanasha, New York.
Corey, E. J., Becker, K. B. and Varma, R. K. (1972) *J. Am. Chem. Soc.*, **94**, 8616.
Cram, D. J. (1976) in *Applications of Biochemical Systems in Organic Chemistry, Techniques of Organic Chemistry Series* (eds J. B. Jones, C. J. Sih and D. Perlman), Wiley Interscience, New York, Vol. 2, p. 815.
Cram, D. J. and Sogah, G. D. Y. (1981) *J. Chem. Soc., Chem. Commun.*, 625.
Deol, B. S., Ridley, D. D. and Simpson, G. W. (1976) *Aust. J. Chem.*, **29**, 2459.
DiCosimo, R., Wong, C.-H., Daniels, L. and Whitesides, G. M. (1981) *J. Org. Chem.*, **46**, 4622.
Dodds, D. R. and Jones, J. B. (1982) *J. Chem. Soc., Chem. Commun.*, 1080.
Fleming, I. (1973) *Selected Organic Syntheses*, Wiley, London, p. 36.
Fronza, G., Fuganti, C. and Graselli, P. (1980) *J. Chem. Soc., Chem. Commun.*, 442.
Fuganti, C. and Graselli, P. (1979) *J. Chem. Soc., Chem. Commun.*, 995.
Fuganti, C. and Graselli, P. (1982) *J. Chem. Soc., Chem. Commun.*, 205.
Guthrie, J. P. (1976) in *Applications of Biochemical Systems in Organic Chemistry, Techniques of Organic Chemistry Series* (eds J. B. Jones, C. J. Sih and D. Perlman), Wiley Interscience, New York, Vol. 2, p. 627.
Hill, R. L. and Teipel, J. W. (1971) in *The Enzymes* 3rd edn (ed. P. D. Boyer), Academic Press, New York, Vol. 5, p. 556.
Hirschbein, B. L. and Whitesides, G. M. (1982) *J. Am. Chem. Soc.*, **104**, 4458.
Iguchi, S. Nakai, H., Hayashi, M. and Yamamoto, H. (1979) *J. Org. Chem.*, **44**, 1363.

Jones, J. B. (1980) in *Enzymic and Nonenzymic Catalysis* (eds P. Dunnil, A. Wiseman and N. Blakebrough), Ellis and Horwood, Chichester, p. 54.

Jones, J. B. and Beck, J. F. (1976) in *Applications of Biochemical Systems in Organic Chemistry, Techniques of Organic Chemistry Series* (eds J. B. Jones, C. J. Sih and D. Perlman), Wiley Interscience, New York, p. 107.

Jones, J. B. and Davies, J. (1981) *J. Am. Chem. Soc.*, **101**, 5405.

Jones, J. B. and Jakovac, I. J. (1982) *Can. J. Chem.*, **60**, 19.

Jones, J. B. and Lok, K. P. (1979) *Can. J. Chem.*, **57**, 2533.

Jones, J. B. and Schwartz, H. M. (1981) *Can. J. Chem.*, **59**, 1574.

Jones, J. B. and Schwartz, H. M. (1982) *Can. J. Chem.*, **60**, 1030.

Jones, J. B. and Taylor, K. E. (1976) *Can. J. Chem.*, **54**, 2969; *J. Am. Chem. Soc.*, **98**, 5689.

Jones, J. B., Finch, M. A. W. and Jakovac, I. J. (1982) *Can. J. Chem.*, **60**, 2007.

Jones, J. B., Sih, C. J. and Perlman, D. (eds) (1976) *Applications of Biochemical Systems in Organic Chemistry, Techniques of Organic Chemistry Series*, Wiley Interscience, New York.

Julia, S., Masana, J. and Vega, J. C. (1980) *Angew. Chem. Int. Edn. Engl.*, **19**, 929.

Kawai, K.-I., Imuta, M. and Ziffer, H. (1981) *Tetrahedron Lett.*, **22**, 2527.

Klibanov, A. M., Berman, Z., and Alberti, B. N. (1981) *J. Am. Chem. Soc.*, **103**, 6263.

Kluender, H., Huang, F.-C., Fritzberg, A., Schnoes, H., Sih, C. J., Fawcett, P. and Abraham, E. P. (1973) *J. Am. Chem. Soc.*, **95**, 6149.

Krishnamurthy, S. (1974) *Aldrichim. Acta*, **7**, 55.

Krishnamurthy, S. (1981) *J. Org. Chem.*, **46**, 4629.

Leuchs, H.-J., Lewis, J. M., Rios-Mercadillo, V. M. and Whitesides, G. M. (1979) *J. Am. Chem. Soc.*, **101**, 5830.

Lewis, J. M., Haynie, S. L. and Whitesides, G. M. (1979) *J. Org. Chem.*, **44**, 864.

Link, C. M., Jansen, D. K. and Sukenik, C. N. (1980) *J. Am. Chem. Soc.*, **102**, 7798.

Menger, F. M. (1979) *Acc. Chem. Res.*, **12**, 111.

Menger, F. M., Takeshita, M. and Chow, J. F. (1981) *J. Am. Chem. Soc.*, **103**, 5938.

Midland, M. M., Tramontano, A. Zderic, S. A. and Greer, S. (1979) *J. Am. Chem. Soc.*, **101**, 2352.

Mosbach, K. and Larsson, P. M. (1970) *Biotechnol. Bioeng.*, **12**, 19.

Neidlemann, S. L., Amon, W. F., Jr. and Geigert (1981) *US Patent* 4 247 641, *Chem. Abstr.*, **94**, 190 337.

Norman, R. O. C. (1982) in *The Chemical Industry* (eds D. Sharp and T. F. West), Ellis and Horwood, Chichester, p. 347.

Noyori, R. Tomino, I. and Tanimoto, Y. (1979) *J. Am. Chem. Soc.*, **101**, 3129, 5843.

Oku, J. and Inoue, S. (1981) *J. Chem. Soc., Chem. Commun.*, 229.

Onyiriuka, S. O. and Suckling, C. J. (1982) *J. Chem. Soc., Chem. Commun.*, 833.

Perlman, D. (1976) in *Applications of Biological Systems to Organic Chemistry, Techniques of Organic Chemistry Series* (eds J. B. Jones, C. J. Sih and D. Perlman), Wiley Interscience, New York, Vol. 1, p 47.

Pollak, A., Baughn, R. L., Adalstensson, O. and Whitesides, G. M. (1978) *J. Am. Chem. Soc.*, **100**, 304.

Pollak, A., Baughn, R. L. and Whitesides, G. M. (1977) *J. Am. Chem. Soc.*, **99**, 2366.

Prelog, V. (1964) *Pure. Appl. Chem.*, **9**, 126.

Reid, T. W., Stein, T. P. and Fahrney, D. (1967) *J. Am. Chem. Soc.*, **89**, 7125.

Rideout, D. C. and Breslow, R. (1980) *J. Am. Chem. Soc.*, **102**, 7816.

Ruddleston, J. F. and Stewart, A. (1981) *J. Chem. Res. (S)*, 378.

Scott, A. I. (1976) in *Applications of Biochemical Systems in Organic Chemistry,*

Techniques of Organic Chemistry Series (eds J. B. Jones, C. J. Sih and D. Perlman), Wiley Interscience, New York, Vol. 2, p. 555.

Sharpless, K. B. and Verhoeven, T. R. (1979) *Aldrichim. Acta.*, **13**, 13.

Sih, C. J. and Rosazza, J. P. (1976) in *Applications of Biochemical Systems to Organic Chemistry, Techniques in Organic Chemistry Series* (eds J. B. Jones, C. J. Sih and D. Perlman), Wiley Interscience, New York, Vol. 1, p. 69.

Sih, C. J., Heather, H. B., Sood, R., Price, P., Peruzzotti, B., Hsu-Lee, L. F. and Lee, S. S. (1975) *J. Am. Chem. Soc.*, **97**, 865.

Simon, H., Gunther, H., Bader, J. and Tischer, W. (1981) *Angew. Chem. Int. Edn. Engl.*, **20**, 861.

Slama, J. T., Oruganti, S. R. and Kaiser, E. T. (1981) *J. Am. Chem. Soc.*, **103**, 6211.

Suckling, C. J. (1977) *Chem. Soc. Rev.*, **7**, 215.

Suckling, C. J. (1981) *Ind. Eng. Chem. Prod. Res. Dev.*, **20**, 434.

Suckling, C. J. (1982) *J. Chem. Soc., Chem. Commun*, 661.

Suckling, C. J. and Suckling, K. E. (1974) *Chem. Soc. Rev.*, **4**, 387.

Suckling, C. J., Suckling, C. W. and Suckling, K. E. (1978) *Chemistry through Models*, Cambridge University Press, Cambridge, pp. 203–210.

Sutherland, I. O. (1982) in *The Chemical Industry* (eds D. Sharp and T. F. West), Ellis and Horwood, Chichester, p. 421.

Szabo, W. A. and Lee, S. S. (1975) *J. Am. Chem. Soc.*, **97**, 865.

Szajewski, R. P. and Whitesides, G. M. (1980) *J. Am. Chem. Soc.*, **102**, 2011.

Tabushi, I. (1982) *Acc. Chem. Res.*, **15**, 66.

Tabushi, I., Yamamura, K., Fujita, K. and Kawakubo, H. (1979) *J. Am. Chem. Soc.*, **101**, 1019.

Trost, B. M. and Klun, T. P. (1979) *J. Am. Chem. Soc.*, **101**, 6756.

Trost, B. M. and Klun, T. P. (1981) *J. Am. Chem. Soc.*, **103**, 1864.

Whitesides, G. M. (1976) in *Applications of Biochemical Systems in Organic Chemistry, Techniques of Organic Chemistry Series* (eds J. B. Jones, C. J. Sih and D. Perlman), Wiley Interscience, New York, Vol. 2, Chapter 7.

Whitlock, H. W., Jr. (1976) in *Applications of Biochemical Systems in Organic Chemistry, Techniques of Organic Chemistry Series* (eds J. B. Jones, C. J. Sih and D. Perlman), Wiley Interscience, New York, Vol. 2, p. 1045.

Wilson, M. E. and Whitesides, G. M. (1978) *J. Am. Chem. Soc.*, **100**, 505.

Wong, C.-H. and Whitesides, G. M. (1981) *J. Am. Chem. Soc.*, **103**, 4890.

Wynberg, H. and Marsmann, B. (1980) *J. Org. Chem.*, **45**, 158.

5 | Enzymes as targets for drug design

Barrie Hesp and Alvin K. Willard

5.1 INTRODUCTION

Alfred Burger (1980) dates the modern era of Medicinal Chemistry from the discovery of the anti-bacterial properties of sulphanilamide in 1936. The subsequent observation that sulphanilamide antagonized the incorporation of *p*-aminobenzoic acid into folic acid marked the beginning of a shift in the medicinal chemist's attention from endogenous metabolites *per se* to the biochemical processes which control the flux of these metabolites in the organism. Many of these important biochemical processes are the enzyme-catalysed reactions of biosynthesis and catabolism. Most others are receptor-mediated processes in which interaction of an endogenous ligand with a receptor in a cell membrane initiates a complex sequence of biochemical events. From an impromptu survey of the research topics in our own organization our portfolio of projects appears to be based almost equally on receptor and enzyme mechanisms. The balance is unlikely to be greatly different elsewhere.

In this chapter we shall be concerned only with enzymes as targets for drug design. For the organic chemist, raised on a diet of reaction mechanisms, enzymes offer some advantages over receptors since our understanding of receptor organization and mechanism lags far behind the detailed information available for many enzyme-catalysed reactions. No detailed X-ray structural data are available for receptors; we can only speculate as to their geometry on the basis of our knowledge of the three-dimensional structures of molecules which bind to them. Enzymes have often been purified and characterized and fine details of structure and mechanism are sometimes available from X-ray studies, though we hasten to add that such a wealth of information is the exception rather than the rule. Even when fact gives way to speculation, enzyme mechanisms conform to the basic rules of organic chemistry and some knowledge of the mechanism of the uncatalysed re-action is often a starting point for the design of inhibitors, as was discussed in Chapter 2.

There are numerous enzymes involved in life processes but many, indeed perhaps the majority, are unsuitable targets for drug intervention. Most of today's clinically useful drugs have resulted from research conducted in pharmaceutical companies. Once a new drug has been identified, the development phase is of the order of $7-10$ years and the cost $\$50-70$ million in the US alone. With such economics, selection of the target enzyme is of crucial importance.

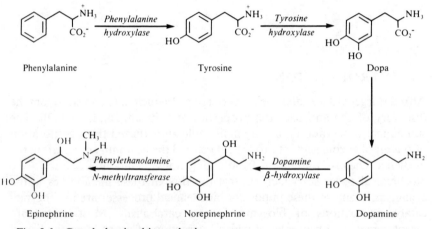

Fig. 5.1   Catecholamine biosynthesis.

Consider for example the enzymes of the catecholamine biosynthetic pathway (Fig. 5.1). Inhibitors of tyrosine hydroxylase and dopamine (3,4-dihydroxyphenethylamine) β-hydroxylase lower blood pressure, primarily by reducing synthesis of the neurotransmitter, norepinephrine (nor-adrenaline). However, such substances are unacceptable as drugs for anti-hypertensive therapy because of their intrinsic lack of selectivity in control of adrenergic function. They interfere with norepinephrine synthesis in blanket fashion, leading to unacceptable side effects, for example orthostatic hypo-tension, resulting from blockade of cardiovascular reflexes. Good ideas for inhibitors of tyrosine hydroxylase are unlikely to receive serious attention in the present-day pharmaceuticals industry for this reason. There are alterna-tive, selective ways of controlling blood pressure via adrenergic mechanisms, for example by blockade of β-adrenergic receptor systems.

In contrast, phenylethanolamine *N*-methyltransferase (PNMT), the terminal step in the biosynthesis of epinephrine (adrenaline), is receiving increasing attention as evidence grows for a role of epinephrine in the central control of blood pressure and in the peripheral physiological response to stress (Bondinell *et al.*, 1980; Fuller, 1982). There is no guarantee that inhibitors of PNMT will not produce side effects; indeed one can speculate as to what they might be. But at least in principle inhibition of the final step in the biosynthesis of epinephrine will have less profound effects on systems regulated by dopamine and norepinephrine than will inhibition of tyrosine hydroxylase.

Our intent in this chapter is to use five case studies to exemplify the evolution of the medicinal chemist's appreciation of enzymes as targets for drug design. Each concerns drugs of established importance or research topics of great current interest. They span the range from drugs developed prior to any knowledge of their target enzymes, to today's attempts at *ab initio* design. Each is a rich source of information for the aspiring medicinal chemist, though this will be neither a comprehensive review of enzyme inhibitors nor shall we address but a few disease areas. Finally, we draw attention to the potential inherent in modern computer graphics and related techniques, particularly in conjunction with new insights into the energetics and mechanisms of enzyme-catalysed reactions.

5.2 CASE STUDIES IN DRUG DISCOVERY

5.2.1 *Inhibitors of dihydrofolate reductase*

Several clinically important drugs produce their therapeutic effects through inhibition of the enzyme dihydrofolate reductase. Dihydrofolate reductase catalyses the reduction of dihydrofolate (*5.1*) to tetrahydrofolate (*5.2*) via transfer of hydride from NADPH (Fig. 5.2). A number of mammalian and microbial cellular processes utilize tetrahydrofolate derivatives as cofactors in one-carbon transfer reactions but the series of reactions coupled to thymidylate synthetase is especially attractive as a target for therapeutic intervention since cellular DNA synthesis cannot be maintained in the absence of functioning thymidylate synthetase. Since transfer of the one-carbon moiety from 5,10-methylenetetrahydrofolate (*5.3*) to deoxyuridylate occurs concomitantly with oxidation of the cofactor to the dihydrofolate level, thymidylate synthetase is critically dependent upon continual regeneration of tetrahydrofolate (*5.2*) from dihydrofolate (*5.1*) by dihydrofolate reductase (Hitchings and Roth, 1980). Thus, inhibitors of dihydrofolate reductase effectively prevent cell division throughout the microbial and animal world.

Before considering in more depth the mechanism of action of dihydrofolate reductase inhibitors we shall briefly discuss their origins. The search

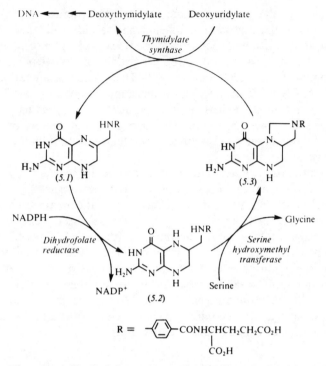

Fig. 5.2 The role of dihydrofolate reductase in the biosynthesis of deoxyribonucleic acid (DNA).

for anti-malarial agents proceeded apace throughout World War II. Curd and colleagues at ICI based their synthetic efforts on pyrimidine analogues because of the key role of pyrimidines in cell metabolism and the known anti-malarial activity of sulphonamides, a class which included pyrimidine derivatives. Synthetic exploration of pyrimidines quickly led to a new series of anti-malarial agents, typified by the diaminopyrimidine derivative (5.5), which was conceived as incorporating the aniline moiety of the sulphona-mides, the basic side chain of earlier alkaloid anti-malarials, and the pyrim-idine nucleus. Further speculative modification of such structures was aimed at the synthesis of compounds resembling the various tautomeric forms available to these heterocyclic amines. The presence of a ring system was not considered to be obligatory and indeed, the compound which subsequently became a successful drug, proguanil (5.6), was an acyclic biguanide. Proguanil is an early sample of a prodrug, since it is active only through metabolism to cycloguanil (5.7a), the first member of the dihydrotriazine series (5.7).

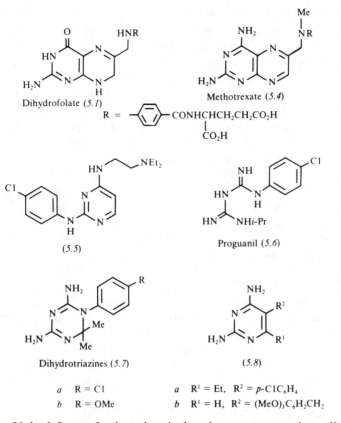

Dihydrofolate (5.1)

Methotrexate (5.4)

R = —〈benzene〉—CONHCHCH₂CH₂CO₂H with CO₂H

$$R = \text{—C}_6\text{H}_4\text{—CONHCHCH}_2\text{CH}_2\text{CO}_2\text{H}$$
$$\qquad\qquad\qquad\ \ |$$
$$\qquad\qquad\qquad\ \ \text{CO}_2\text{H}$$

(5.5)

Proguanil (5.6)

Dihydrotriazines (5.7)

(5.8)

a   R = Cl   a   R¹ = Et, R² = p-ClC₆H₄
b   R = OMe   b   R¹ = H,   R² = (MeO)₃C₆H₂CH₂

$$a \quad R = Cl$$
$$b \quad R = OMe$$

$$a \quad R^1 = Et, \quad R^2 = p\text{-ClC}_6\text{H}_4$$
$$b \quad R^1 = H, \quad R^2 = (\text{MeO})_3\text{C}_6\text{H}_2\text{CH}_2$$

In the United States further chemical endeavours centred on diamino-pyrimidines (5.8) by the Burroughs Wellcome group led initially to pyrimethamine (5.8a) and then trimethoprim (5.8b). Also in the 1940s, a mechanistic approach to new anti-bacterial agents conducted by scientists at American Cyanamid, aimed at the design of antagonists of the known bacterial growth factor, folic acid, led directly to another diamino heterocycle, methotrexate (5.4).

Though not fully appreciated at the time, all of the active compounds described above work by inhibiting the enzyme dihydrofolate reductase. In the 1980s much emphasis is rightly given to rational drug design and the medicinal chemist's efforts are critiqued by his peer group with this criterion in mind. Similar assessments of earlier research are only legitimate if due consideration is also given to the constraints of the day. By these yardsticks the efforts which led to the dihydrofolate reductase inhibitors are truly excellent examples of rational drug design.

Because of the ubiquitous distribution of the enzyme, the therapeutic value of dihydrofolate reductase inhibitors would be limited but for the selectivity differences observed for the various chemical subtypes. Some members of the triazine and pyrimidine classes of inhibitors display remarkable specificites for the enzymes from different species. Pyrimethamine (5.8a) is extremely effective as an inhibitor of the enzyme from the malaria parasite and is 2000–5000 times less potent against bacterial and mammalian enzyme: hence its clinical utility in malaria. Trimethoprim (5.8b) is most potent versus bacterial enzymes, being some 14 times less effective against enzyme from the malaria parasite and essentially inactive versus mammalian enzyme. This selectivity for the bacterial enzyme underlies the efficacy and widespread anti-bacterial usage of this drug, particularly in synergistic combination with a sulphonamide – a class of drug which interferes with bacterial folic acid synthesis at a different step, namely the incorporation of p-aminobenzoic acid. In contrast, some of the dihydrotriazines (5.7) are 1000 times more potent on mammalian than bacterial enzymes. Methotrexate (5.4) is non-selective in vitro, inhibiting enzyme from vertebrates and bacteria almost equally. Despite methotrexate's high affinity for microbial enzymes it is of no value clinically as an anti-microbial agent since protozoa and bacteria, which synthesize their own folate, generally lack the uptake systems present in mammalian cells for folate and its congeners. Hence methotrexate (5.4), a structural analogue of folic acid, does not penetrate efficiently into microbial cells but readily enters mammalian cells. Its main clinical use is to kill cancerous cells; the rationale being that such cells are rapidly proliferating and therefore greatly dependent on DNA synthesis, and somewhat more susceptible to DNA synthesis inhibitors than are normal cells.

Hitchings et al. have commented that species differentiation may be more difficult to achieve in enzyme inhibitors which are close structural analogues of the substrate than in inhibitors from more distant chemical classes (Hitchings and Roth, 1980; Hitchings and Smith, 1980).

A complete account of dihydrofolate reductase must delineate the mechanism of action of the enzyme, the interactions between protein and cofactors, substrate or inhibitors, and must account for the interspecies differences in the selectivities of the enzyme for the various inhibitors. Studies with this group of enzymes exemplify the power of modern biochemical, physical organic, spectroscopic and X-ray-crystallographic techniques in elucidating the fine details of enzyme mechanisms. The molecular basis for the interspecies selectivity differences are less well understood.

Structure determinations have been made by X-ray methods for examples of each of the three major classes of inhibitors bound to the enzyme. Two of these studies are of ternary complexes, viz. bacterial enzyme (L. casei)–NADPH–methotrexate (Matthews et al., 1978) and avian enzyme–NADPH–dihydrotriazine (Volz et al., 1982); the third is of the binary complex, bacterial enzyme (E. coli)–trimethoprim (Baker et al., 1981).

Despite the differences between the enzymes all three classes of inhibitor interact with the protein residues in the active site in a remarkably similar manner.

Methotrexate (*5.4*) binds in an open conformation in a deep cavity which cuts across one face of the enzyme. The pteridine ring is almost perpendicular to the aromatic ring of the *p*-aminobenzoyl group. There are 17 interactions between methotrexate and the enzyme, ten of which involve the pteridine ring. The charge interaction between Asp-26 (*L. casei*) and N_1 of the pteridine appears to be of key importance in the binding of the pteridine fragment. The dihydrotriazine inhibitor binds to the avian enzyme with the triazine and methoxyphenyl rings in corresponding positions to the pyrimidine and aminoalkylpyrazine fragments, respectively, in the enzyme–methotrexate complex.

With the proviso that methotrexate (*5.4*) is an inhibitor, not a substrate, the geometric relationship between NADPH and the pteridine ring of methotrexate is consistent with the experimental evidence that transfer of hydride is from the *A* face of the reduced nicotinamide ring (Fig. 5.3(a)).

Each class of inhibitor is unprotonated in the unbound state at neutral pH (unbound methotrexate N_1 $pK_a = 5.73$) but is protonated at N_1, with $pK_a \geq 10$, in the bound form (Volz *et al.*, 1982; Cocco *et al.*, 1981; Ozaki *et al.*, 1981; Roberts *et al.*, 1981). For the *E. coli* enzyme the charge interaction

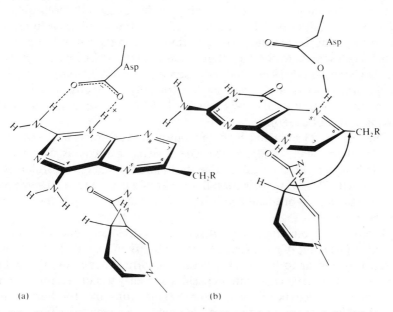

(a)   (b)

Fig. 5.3 Orientation of methotrexate (a) and dihydrofolate (b) in the active site of dihydrofolate reductase.

is between N_1 of trimethoprim and Asp-27, the analogous interaction for the avian enzyme is between N_1 of the triazine and Glu-30. The available evidence is that dihydrofolate is unprotonated, either free at physiological pH, or bound to the enzyme. Recently Cocco *et al.* (1981) have reasoned that the greater than 1000-fold difference in affinities of methotrexate and dihydrofolate for the enzyme is almost entirely due to the high pK_a of N_1 in bound methotrexate. This extends Baker's earlier suggestion that the greater affinity of methotrexate stems from the increased basicity of a diaminopyrimidine *versus* a 2-amino-4-oxopyrimidine.

At first inspection the structural similarity of methotrexate (*5.4*) to the natural substrate, dihydrofolate (*5.1*), is immediately evident and one might expect that the substrate and the inhibitor would bind at the active site in very similar, if not identical, conformations. Matthews *et al.* (1978) have discussed the stereochemical implication inherent in such an assumption, namely that transfer of hydride from C_4 of the dihydronicotinamide ring to C_6 of the dihydropyrazine ring would lead to the enantiomer with the R absolute configuration at C_6. An alternative, but less obvious, binding mode was suggested in which the pteridine ring in dihydrofolate is rotated 180° to interchange N_1 and N_8 with C_4 and N_5, respectively, leaving the side chain from N_{10} essentially unchanged: such a binding mode would lead to tetrahydrofolate with the S absolute configuration at C_6. An X-ray analysis of the enzyme–substrate complex has not yet been reported but compelling evidence for the latter, 'inverted', binding mode has recently been obtained. Thus X-ray analysis of 5,10-methenyl-5,6,7,8-tetrahydrofolic acid bromide has shown that the absolute configuration at C_6 in the natural isomer is R, which corresponds to the S configuration for tetrahydrofolate (*5.2*) (Fontecilla-Camps *et al.*, 1979); and reduction of 6-methyl-7,8-dihydropterin by dihydrofolate reductase gave a single tetrahydro derivative, for which the absolute configuration at C_6 was shown to be S by correlation with (S)-alanine (Armarego *et al.*, 1980).

The proposed productive binding mode of dihydrofolate to the enzyme is shown schematically in Fig. 5.3(b). The aspartate carboxyl is shown unionized in line with current thinking that the bound form of dihydrofolate is unprotonated. Clearly, there is formal transfer of a proton to N_5 at some point on the reaction co-ordinate and it is possible that aspartate is the source of the proton.

The different binding modes of such closely related analogues as methotrexate (*5.4*) and dihydrofolate (*5.1*) exemplify the ramifications of what frequently appear to be trivial structural modifications in the interactions between small, drug candidate molecules and enzymes or receptors. In this particular instance the minor structural changes of 4-oxo for 4-amino both switches the binding mode and results in greatly enhanced affinity for the active site.

The heterogeneity of dihydrofolate reductases from different species with respect to binding affinities for the three major classes of inhibitor was referred to earlier. An understanding of such differences constitutes a formidable challenge. Some clear differences in the constitution of the enzymes are known. For example, bacterial enzymes have molecular weights of approx. 18 000 and show relatively little sequence identities with enzymes of animal origin or from other species of bacteria. Animal reductases generally contain 30 or so fewer amino acids than those from bacteria, and are highly homologous. Protozoan enzymes are huge (molecular weight approx. 180 000) in comparison.

From the recent literature it is evident that much effort continues to be devoted to the design of inhibitors of dihydrofolate reductase. In this context the detailed structural and mechanistic information which has emerged in recent years will be of great assistance. One particularly interesting study was published recently by scientists at Burroughs-Wellcome (Kuyper *et al.*, 1982). From a consideration of interactions observed in the active site from X-ray studies they were able to design analogues of trimethoprim with greatly enhanced potency. It remains uncertain, however, whether the deficiencies in the dihydrofolate reductase inhibitors currently used in medicine are sufficient to justify the cost of developing new drugs acting by the same mechanism.

5.2.2 Non-steroidal anti-inflammatory agents

The history of medicinal chemistry abounds with examples of drug searches in which recognition that the target at the molecular level was an enzyme came long after the drug was discovered. In the previous section we saw how the anti-bacterial properties of compounds designed as anti-metabolites were ultimately related to inhibition of dihydrofolate reductase. In many other areas drug design was based on animal models of a disease, which only in retrospect were recognized to be dependent on a particular enzyme or biosynthetic pathway. Often, a clear understanding of the biochemical mechanism of a particular disease state only became possible after the drug was available. The discovery of a class of drugs known as 'non-steroidal anti-inflammatory drugs' (NSAIDs) is an excellent example of this scenario.

Interesting insights into medicinal chemist's thinking can be gained by a brief excursion through the history of the NSAIDs. These drugs arose through research programmes aimed at alleviation of rheumatoid arthritis.

Aspirin (*5.9*) had been in use as an anti-inflammatory drug for half a century or so before the powerful anti-inflammatory steroids were introduced in the 1950s. The steroids are highly effective drugs but their usefulness is severely limited by their propensity to produce hormonal and metabolic side

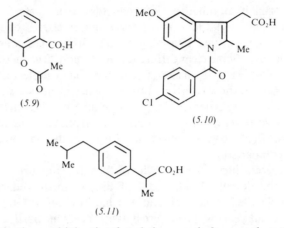

(5.9)

(5.10)

(5.11)

effects − limitations which stimulated the search for novel structures with anti-inflammatory activity.

Chemists at Merck (Shen and Winter, 1977; Shen, 1981) in the early 1960s were guided by two observations. The first was the potent inflammatory effect of serotonin (5-hydroxytryptamine) and the second, an observation of abnormal tryptophan metabolism in rheumatic patients. We know now that neither of these factors is important for the activity of the drugs which evolved from the programme, but both served to focus attention on indoles. These efforts culminated in the discovery of indomethacin (5.10) in 1961. An intensive search ensued for the mechanism of action of this drug, and many others which quickly followed, for example ibuprofen (5.11), but it was not until ten years later that Vane (1971) announced that both indomethacin and aspirin inhibit the biosynthesis of prostaglandins, some of which are important mediators of inflammation. Evidence was soon amassed to correlate this *in vitro* observation with the *in vivo* and clinical properties of the compounds.

The prostaglandins represent some of the members of a cascade of products which arise from metabolism of arachidonic acid (5.12) (Fig. 5.4). Their production is controlled by the enzyme cyclo-oxygenase whose ultimate product is the cyclic endoperoxide PGG_2 (5.13). Although several mammalian cyclo-oxygenases have been purified to homogeneity (Samuelsson et al., 1978), most of the details of this highly complex enzymic transformation have not been elucidated. In the presence of haem the enzyme catalyses the cyclo-oxygenase reaction which converts arachidonic acid (5.12) to PGG_2 (5.13). Additionally, in the presence of tryptophan further cleavage of the hydroperoxy group of PGG_2 occurs to produce PGH_2 (5.14). The ratio of cyclo-oxygenase and peroxidase activities does not change during purification, but experiments with purified enzyme demonstrate that aspirin and indomethacin inhibit the cyclo-oxygenase activity, but not the peroxidase activity.

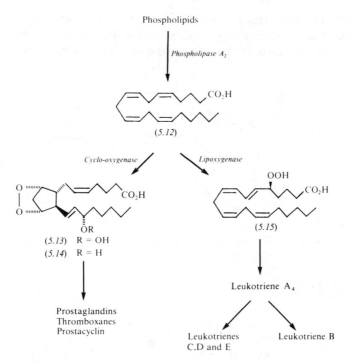

Fig. 5.4 Dual pathways of arachidonic acid metabolism.

It is appropriate here to recall our earlier discussion of the importance of selection of the proper enzyme. Cyclo-oxygenase was discovered after the NSAIDs. If, however, enzyme selection had preceded the drug search, many scientists would have considered cyclo-oxygenase to be an inappropriate point at which to interrupt arachidonic acid metabolism. Prostaglandins are present in almost every tissue and play a multitude of biological roles in the organism. For example, prostacyclin and thromboxane provide the critical balance in platelet—vessel wall interactions involved in clotting and thrombus formation. Indeed, some of the side effects of the NSAIDs can be attributed to reduction of levels of arachidonic acid metabolites in biological roles other than inflammation. Remarkably, such concerns, had they existed, would have been largely unfounded since the NSAIDs are among the most widely used drugs in medicine today (ibuprofen was the third largest selling drug in the US in 1981).

The transformation of arachidonic acid to PGG_2 is a complex process but a range of studies has uncovered enough of the mechanistic and stereochemical details of the cyclo-oxygenase reaction to allow scientists to postulate mechanisms for the reaction (Samuelsson *et al.*, 1978; Walsh, 1981). Among others, Gund and Shen (1977) have proposed a plausible mechanism

(Fig. 5.5). The first step is presumably an allylic hydroperoxylation of the Δ^{11}-double bond with rate-determining pro-*S* hydrogen atom abstraction at C-13. The enzyme contains both haem and non-haem iron, suggesting that

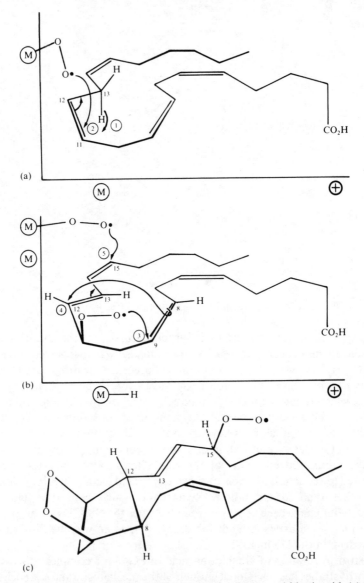

Fig. 5.5 Postulated mechanism of PGG$_2$ formation from arachidonic acid. (a) Allylic hydroperoxidation of the 11,12-double bond. (b) Cyclization and further oxidation of allylic hydroperoxy radical. (c) Conformation of PGG$_2$ in the active site (Gund and Shen, 1977; Copyright 1977, American Chemical Society; used with permission).

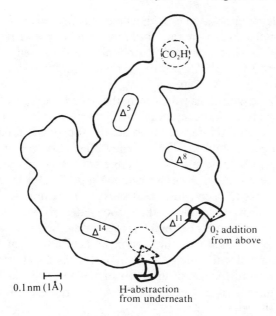

Fig. 5.6 Hypothetical active site of cyclo-oxygenase (Gund and Shen, 1977; Copyright 1977, American Chemical Society; reprinted with permission).

both O_2 activation and hydrogen removal could be metal-mediated. This intermediate hydroperoxide can escape but normally further oxidation in the same active site leads to PGG_2. This requires closure of the peroxy radical on the Δ^8-double bond at C-9 and conrotatory ring closure between C-8 and C-12. The resulting allylic radical adds another molecule of oxygen at C-15. Both chemiluminescence (singlet oxygen decay?) and electron paramagnetic resonance (EPR) signals have been observed during cyclo-oxygenase catalysis.

Following the discovery of indomethacin, two decades of medicinal chemistry produced a myriad of NSAIDs, most of which are aromatic acetic or propionic acids. These compounds led to, or benefited from, a variety of hypothetical models of a 'receptor' for the NSAIDs. One of these models was proposed by Gund and Shen (1977), in an early attempt by medicinal chemists to take advantage of conformational analysis by computer-assisted molecular modelling. They incorporated the mechanistic information summarized above and of structural information from a range of NSAIDs as well as from arachidonic acid itself. The model which they proposed as capable of accommodating most of the structural information at hand consisted of a planar hydrophobic area with a carboxylic acid-binding site below the hydrophobic plane (Fig. 5.6) The Δ^5 and Δ^8 double bonds, as well as the carboxylate of arachidonic acid, are accepted by this region. Further they proposed a hydrophobic groove which accommodates the Δ^{11} and Δ^{14}

double bonds of arachidonic acid where the oxidative transformations occur. Indomethacin can interact with this model of the active site with the indole ring binding in the planar hydrophobic site. The phenyl ring is accommodated by the hydrophobic groove, perhaps with the p-chloro substituent interacting with the metal centre.

Many of the assumptions which form the basis for design of models such as this are open to criticism, especially with an enzyme like cyclo-oxygenase for which no structural information exists and for which the catalytic mechanism is uniquely complex. NSAIDs may bind only in a fraction of the arachidonic acid-binding sites while taking advantage of sites that are not occupied by the substrate. Drugs which upon superficial examination appear similar may prove to occupy different sites. Even assumptions that appear well founded, such as binding of the carboxylates of arachidonic acid and the NSAIDs at the same site, might prove wrong. Scientists at Fisons, for example, have proposed that the carboxylic acid groups of the NSAIDs, including indomethacin, mimic the 11-hydroperoxy group and co-ordinate to the metal centre (Appleton and Brown, 1979). They suggest that this model accommodates a variety of NSAID structures better than the Gund–Shen model.

Models such as these are interesting as after-the-fact rationalizations of structure activity information; they also provide the medicinal chemist with a conceptual framework on which to design future cyclo-oxygenase inhibitors. However, models of the cyclo-oxygenase active site are especially interesting since they may also provide the medicinal chemist with the rare opportunity to lift a model, almost intact, and apply it to another enzyme, 5-lipoxygenase.

5-Lipoxygenase stands at the head of a second major pathway of arachidonic acid metabolism which has been elucidated recently. This 'linear' pathway produces a family of metabolites known as the leukotrienes (Fig. 5.4) which are now recognized as important mediators of asthma and other allergic reactions (Borgeat and Sirois, 1981). Some of the products of this pathway, especially leukotriene B_4 (LTB$_4$), may play a critical synergistic role in the prostaglandin-mediated inflammatory response as well. It may be just as important and beneficial therapy for inflammation to inhibit biosynthesis of the leukotrienes.

The oxidation of arachidonic acid catalysed by 5-lipoxygenase produces 5-hydroperoxy-6,8,11,14-eicosatetraenoic acid *(5.15)* (5-HPETE). This reaction is exactly analogous to the first stage of the cyclo-oxygenase reaction. Although no molecular details have been reported for the 5-lipoxygenase-catalysed reaction or for the active site of 5-lipoxygenase, evidence from cyclo-oxygenase and other lipoxygenase enzymes, our understanding of the electronic and stereochemical requirements of reactions,* and faith in

* See stereoelectronic effects in enzymic catalysis (Chapter 2).

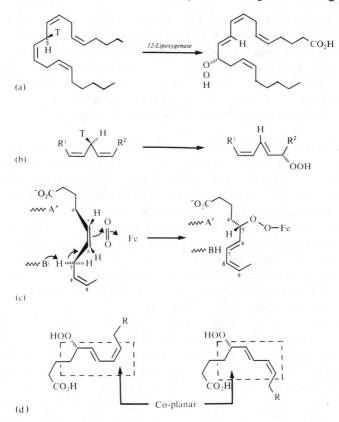

Fig. 5.7 (a) Stereochemical consequences of the 12-lipoxygenase reaction. (b) Stereochemical pathway for generalized lipoxygenase reaction. (c) Hypothetical mechanism for the conversion of arachidonic acid to 5-HPETE by 5-lipoxygenase. (d) Proposed conformations of 5-HPETE in the active site of 5-lipoxygenase.

Nature's economical approach to the design of enzymes allow us to construct a conceptual model of the active site of 5-lipoxygenase. First, the stereochemical details for the related 12-lipoxygenase of platelets and for the 15-lipoxygenase of soybeans have been established by study of the conversion of stereospecifically tritium-labelled arachidonic acid into the corresponding hydroperoxy acids (Hamberg and Hamberg, 1980). In the 12-lipoxygenase pathway the stereochemical findings are described by Fig. 5.7(a). The generalized stereochemical pathway (Fig. 5.7(b)) is followed not only by the lipoxygenase enzymes but also by cyclo-oxygenase in its first allylic hydroperoxylation step discussed above. The lipoxygenases are iron-containing enzymes just as is cyclo-oxygenase. These lines of evidence suggest, but do not prove, that the oxidation mechanisms are similar for all of the enzymes including 5-lipoxygenase.

This assumption allows us to construct a hypothetical pathway for a concerted, iron-mediated 5-lipoxygenase reaction (Fig. 5.7(c)) (not necessarily a paired-electron process). Proper orbital overlap requires co-planarity of atoms 4, 5, 6, 7 and 8. To further weaken the C_7-H bond, the double bonds between atoms 5−6 and 8−9 probably lie in the same plane. These requirements taken together provide us with two hypothetical conformations of the product, 5-HPETE, in the 5-lipoxygenase active site (Fig. 5.7(d)).

Seldom can such detailed structural information be extracted from an enzyme-mediated reaction in the absence of detailed studies of mechanism. This model of the active site of 5-lipoxygenase and conformational restrictions of the substrate arachidonic acid serve as a basis for the rational design of 5-lipoxygenase inhibitors. A planar π-electron system, a carboxylate binding site and an iron centre to which a properly chosen and placed ligand could bind extremely tightly are all familiar from the cyclo-oxygenase active site models discussed above. The challenge to the medicinal chemist is to assimilate these molecular details and to take advantage of the successful rationalization of inhibitor activity in cyclo-oxygenase models to design potent selective 5-lipoxygenase inhibitors.

The cyclo-oxygenase enzyme was discovered long after the drugs had proved therapeutically beneficial. In the lipoxygenase area, on the other hand, the medicinal chemist's target enzyme has been identified before the drug search has begun. Time will tell if the task is easier when the molecular target is recognized at the start of the drug search.

5.2.3   β-Lactam antibiotics

The discovery of penicillin by Fleming, and the pioneering efforts of Florey and colleagues which led to its introduction into medical practice in the early years of the second world war, is a story familiar to most science graduates. The β-lactams now constitute by far the most important class of antibacterial agents and continue to be a focus of attention for the medicinal chemist. (See Chapter 7 for a discussion of the biosynthesis of β-lactams.) No other class of drug has been the subject of such intensive investigations in the history of the pharmaceutical industry. Scarcely a year passes without a major new discovery which shatters earlier views on the structural prerequisites for anti-bacterial activity − compare, for example, the structure of penicillin G (5.16) with those of two newer antibiotics, thienamycin (5.18) and the monocyclic β-lactam, known as a monobactam, SQ 26455 (5.19).

The β-lactams kill bacteria by disrupting bacterial cell wall synthesis, specifically by inhibiting the transpeptidases which cross-link peptidoglycan chains in the final steps of the biosynthetic sequence (Fig. 5.8). The enzyme

formally cleaves a D-alanyl-D-alanine bond at the terminus of one peptidoglycan chain with formation of a new amide linkage between the remaining D-alanine and the terminal glycine of a pentaglycine sequence of a second
peptidoglycan chain (Tipper, 1979; Tomasz, 1979).

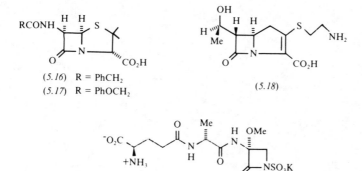

(5.16)   R = PhCH$_2$
(5.17)   R = PhOCH$_2$

(5.18)

(5.19)

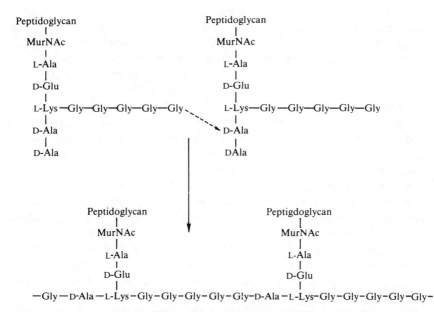

Fig. 5.8 Transpeptidation of two linear peptidoglycan chains in bacterial cell wall
synthesis (Strominger, 1970; Copyright 1970, Academic Press; used with permission).

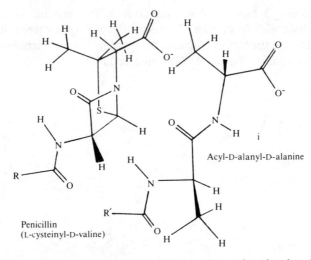

Fig. 5.9 Structural resemblance of penicillin to the acyl-D-alanyl-D-alanine fragment of bacterial cell walls (modified from Strominger, 1970; Copyright 1970,
Academic Press; used with permission).

Tipper and Strominger first proposed that the pencillins are substrate
analogues of the D-alanyl-D-alanine moiety (Fig. 5.9) that is cleaved by the
transpeptidase: more recently the suggestion that penicillin is a transition-
state analogue (see Section 2.8) has received a good deal of attention, it being
argued that the lack of effective conjugation in the amide bond of the β-
lactam is paralleled in the substrate at a point on the reaction co-ordinate.

There were a number of serious shortcomings evident in the early penicillins: they were poorly active when given orally; their spectrum of anti-
bacterial activity was very limited, particularly versus Gram-Negative
bacteria; their *in vivo* half-lives were short owing to rapid renal excretion;
and they were very susceptible to inactivation by bacterial β-lactamases,
enzymes which open the β-lactam ring of the antibiotic. These deficiencies
have been overcome for the most part in the newer agents and this, together
with their excellent safety margin, underlies the commercial success of this
class of antibiotics. Nevertheless, the search for newer, improved agents
continues unabated with emphasis being given to broadening the profile of
activity yet further and improving the stability of the antibiotics to the action
of β-lactamases. The latter remains a serious concern in view of the heterogeneity of the β-lactamases and the propensity for rapid transmission of
plasmid-borne genes throughout the bacterial world. Perhaps the most
intriguing and challenging aspect of the resistance problem is the similarity,
real or apparent, between the enzymes which inactivate the antibiotics and
those which are themselves the targets for the inhibitory action of the drugs.

Studies with radiolabelled β-lactams have shown that the antibiotics bind to one or more penicillin-binding proteins (PBPs) in bacterial cytoplasmic membranes. Seven PBPs have been identified and numbered Ia, Ib, II, III, IV, V and VI in order of decreasing molecular weight. The PBPs are a mixture of transpeptidases and carboxypeptidases, with the latter predominating. The carboxypeptidases differ from the transpeptidases in the nucleophile which serves as acceptor for the penultimate D-alanyl moiety in the peptidoglycan; for the former the acceptor is water, whereas for the latter it is the glycine terminus of a second peptidoglycan. PBPs Ia, Ib, II and III are transpeptidases, some or all of which are essential for cell survival. PBPs IV, V and VI are carboxypeptidases and are not critical for bacterial cell growth. Because of the great technical difficulties in obtaining the transpeptidases, most of our information concerning the inhibitory mechanisms of the penicillins has been obtained from studies with carboxypeptidases: it is assumed that β-lactams inhibit both the transpeptidases and carboxypeptidases by very similar mechanisms.

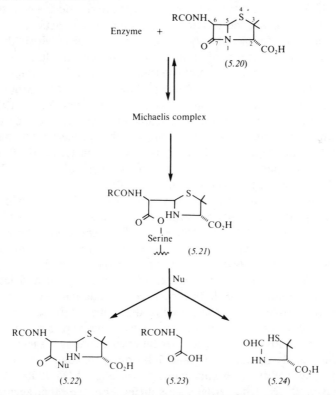

Fig. 5.10   The interaction of penicillins with bacterial β-lactamases and carboxypeptidases: decomposition pathways available to the acyl–enzyme complex. Nu = water, amino acid, alcohol.

The evidence implies that penicillin interacts with lactamases and carboxy-peptidases in a similar fashion (Fig. 5.10). Following initial formation of the Michaelis complex a second complex (*5.21*) forms in which the penicillin has become bound covalently to the enzyme through acylation of a serine residue in the active site (Yocum *et al.*, 1982; Knowles, 1982). In the case of lactamase, enzyme is regenerated by transfer of the penicilloyl moiety to water with formation of a penicilloic acid (*5.22*) (Nu = OH). With the penicillin-sensitive enzymes hydrolytic regeneration of enzyme usually occurs con-comitant with cleavage of the 5,6-bond of the penicillin to yield an acylated glycine (*5.23*) and *N*-formylpenicillamine (*5.24*); though penicilloic acid derivatives (*5.22*) are obtained when the acceptor is an amine or an alcohol (Marquet *et al.*, 1979), no *irreversible* inactivation of the enzyme occurs. There is no evidence to suggest that the products from 5,6-bond fission are derived from an enzyme-bound intermediate distinct from the acyl–enzyme complex (*5.21*). The fragmentation reaction appears to result from the reactivation process, perhaps via mechanisms involved in the release of enzyme-bound D-alanine in the normal substrate reaction. The rate of for-mation and stability of the acylated complex determines lactamase resistance in the context of one class of enzyme and anti-bacterial activity in the context of the other class. The rate of breakdown of the penicilloylated enzyme can be extremely slow for the penicillin-sensitive enzymes; for example the half-life of the benzylpenicillin–carboxypeptidase complex from *Streptomycete* R39 under physiological conditions is 4250 min (Frere *et al.*, 1975). The newer, lactamase-resistant antibiotics are resistant because they are poor substrates for lactamases, yet retain their inhibitor properties for the penicil-lin-sensitive enzymes. Thus, hydrolysis of cefoxitin (*5.25*) by *E. coli* lac-tamase is more than 10^5 times slower than hydrolysis of penicillin V (*5.16*), one of the older penicillins (Fisher *et al.*, 1980).

A new chapter in the β-lactam story began with the isolation of clavulanic acid (*5.26*) from *Streptomyces clavuligerus* by workers at Beecham Pharma-ceuticals in the early 1970s (Brown, 1981). Clavulanic acid differed from all β-lactams studied hitherto in being only a weak anti-bacterial agent but an inhibitor and *irreversible inactivator* of β-lactamases. Good synergistic anti-bacterial effects result from mixtures of clavulanic acid and lactamase-sensitive penicillins. Indeed, one such combination is now marketed in Europe. The discovery of clavulanic acid stimulated the search for other inhibitors from natural sources and through synthesis. As a result, a number of such compounds have emerged in recent years including the semi-synthetics penicillanic acid sulphone (*5.27*) and 6-β-bromopenicillanic acid (*5.28*), and the naturally occurring carbapenems, for example MM 13902 (*5.29*) and thienamycin (*5.18*). The carbapenems differ from the other agents in that they are also good broad-spectrum anti-bacterial agents. For each of the β-lactamase inhibitors the inhibited species is a transformation product,

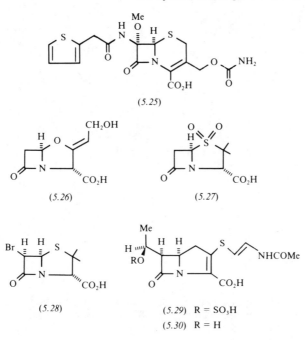

(5.25)

(5.26)   (5.27)

(5.28)   (5.29) R = SO₃H
          (5.30) R = H

or products, of an initial acyl–enzyme complex, rather than the acyl–enzyme complex *per se*. This contrasts with the mechanism by which penicillins inhibit the transpeptidases and carboxypeptidases.

The complex mechanisms by which these compounds inhibit lactamase differ in detail. Knowles (1982) has summarized the various mechanisms as shown in Fig. 5.11. Each inhibitor may be turned over by the normal path to give products derived from the acyl–enzyme intermediate (5.33). Often such products also involve cleavage of the five-membered heterocyclic ring.

The presence of an electron sink X in the sulphones and clavulanic acid ensures diversion of a fraction of the tetrahedral intermediate or acyl–enzyme complex to products (5.34) which may fragment further, or more interestingly, capture a nucleophilic site on the enzyme and thereby inactivate the enzyme irreversibly. For penicillanic acid sulphone approximately 7000 turnovers are required before the enzyme is inactivated. Sulphone analogues which are poorer substrates require fewer turnovers to inactivate the enzyme.

An alternative diversionary path, which does not lead to irreversible inactivation, appears to be available for those carbapenems which can function as lactamase inhibitors (Easton and Knowles, 1982). In such cases the inhibitory species is the Δ¹-pyrroline (5.35), an intermediate which is released from the enzyme much more slowly than the normal acyl–enzyme complex (5.33).

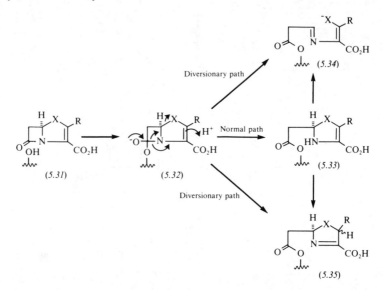

Fig. 5.11 Diversionary pathways that may lead to inhibition or inactivation of β-lactamase (Knowles, 1982; Copyright 1982, Springer Verlag; used with permission).

Within the same chemical series the availability of a diversionary pathway can be influenced profoundly by minor structural modifications. This is perhaps most evident in the case of MM 22382 (*5.30*) and its sulphate ester MM 13902 (*5.29*). The former is rapidly hydrolysed by lactamase with no detectable participation of the diversionary pathway; the latter is a good inhibitor.

The β-lactam series encompasses compounds which are good inhibitors of the transpeptidases but good substrates for lactamase (e.g. penicillin G), good inhibitors of the transpeptidases which are poor substrates for lactamase (e.g. cefoxitin), good inhibitors of both transpeptidases and lactamases (e.g. MM 13902), and finally, compounds which are very poor inhibitors of the transpeptidases but irreversible suicide inactivators of lactamase (e.g. clavulanic acid). What is conspicuous by its absence is a good inhibitor of the transpeptidases which also irreversibly inactives lactamase.

We are far from a complete understanding of the structure–activity relationships in the β-lactam antibiotics, an issue of outstanding importance. It is perhaps because of this lack of understanding that each advance, in terms of overturning earlier notions of such relationships, has come directly from the discovery of new compounds from Nature, rather than from the progressive application of our knowledge of enzyme mechanisms to drug design.

The sparsity of X-ray data on the penicillin-sensitive enzymes, their substrates and inhibitors, contrasts with the situation for dihydrofolate reductase. However, preliminary data for a β-lactam-sensitive bifunctional carboxypeptidase–transpeptidase from *Streptomyces* R61 provide a fore-taste of what is to come (Kelly *et al.*, 1982). The study clearly demonstrates that β-lactam inhibitors and substrate analogues bind to similar regions of the enzyme, consistent with the Strominger hypothesis. A detailed analysis of the X-ray data is eagerly awaited and should add new dimensions to the β-lactam story.

5.2.4   Inhibitors of cholesterol biosynthesis

The β-lactam antibiotics afford an excellent example of the exploitation of natural products as an important source of chemical leads – indeed, valuable drugs. In many other disease areas scientists have also looked to natural products as a rich source of structural variety in their search for drugs. Isolated enzyme assays are particularly amenable to screening of crude extracts or unfractionated fermentation broths because of the high sensitivity and high specificity of the assay. Scores of natural products in each extract or broth can be screened for activity in a single assay and the arduous tasks of isolation and structure determination can be postponed until the level of biological activity merits further effort. This approach to drug discovery has been successfully applied to the search for inhibitors of cholesterol biosynthesis.

Elevated serum cholesterol is a primary risk factor for the development of atherosclerosis and coronary artery disease, the major cause of death in western countries. In humans, at least 50% of total body cholesterol is derived from *de novo* synthesis. Thus, inhibition of cholesterol biosynthesis becomes an attractive target for the pharmaceutical industry.

Interruption of this major metabolic pathway, however, is fraught with potential pitfalls. Cholesterol, the ubiquinones, dolchicol and isopentenyl adenosine, a component of transfer RNA, are all critical products of a common pathway (Fig. 5.12) (Schroepfer, 1981, 1982). Cholesterol itself serves as a substrate for synthesis of the steroid hormones – including oestrogens, testosterone and progesterone. These metabolites play a critical role in maintenance of homoeostasis in the organism. The desirability of interfering with such a fundamental biosynthetic pathway remains open to question. Assuming that this basic objection has been overruled, the choice of which enzyme step to inhibit remains crucial. These concerns cannot be directly assessed, however, until the utility and safety of inhibitors of specific steps are evaluated in humans.

For some cholesterol biosynthesis inhibitors this clinical assessment is already a matter of record. Triparanol (*5.36*) inhibits at a very late stage in the pathway. The drug was introduced for clinical use but withdrawn owing

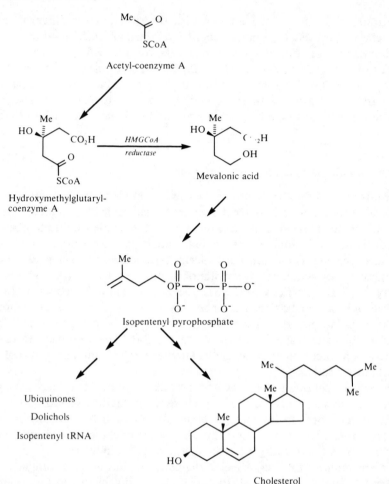

Fig. 5.12 Mevalonate metabolism in mammalian cells.

to serious toxicity which may have been related to a build-up of desmosterol, a biosynthetic intermediate (Laughlin and Carey, 1962; Tobert *et al.*, 1982). This type of toxicity problem justified concerns over the appropriate choice of enzyme for inhibition and served to focus attention much earlier in the pathway where the intermediates, unlike desmosterol, have alternative metabolic options.

The major rate-limiting enzyme in the pathway is 3-hydroxy-3-methyl-glutaryl-coenzyme A reductase (HMG-CoA reductase) (Brown and Rodwell, 1980) which catalyses the reduction of 3-hydroxy-3-methylglutaryl-coenzyme A (*5.37*) to mevalonate (*5.39*) (Fig. 5.13). The enzyme has a unique role in

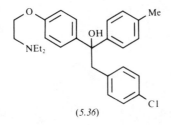

(5.36)

regulation of the biosynthetic pathway. In cultured human fibroblasts, for example, cholesterol from the medium or *de novo* synthesis is required for cell growth. The presence of cholesterol in the medium markedly reduces HMG-CoA reductase activity. Substitution of 7-oxocholesterol, a non-utilizable steroid, for cholesterol severely retards cell growth and lowers reductase activity, presumably by a false feedback mechanism. Growth inhibition is reversed by addition of cholesterol or mevalonate to the medium but not by addition of acetate or hydroxymethylglutarate. These experiments clearly demonstrate that HMG-CoA reductase is the site of feedback regulation of cholesterol synthesis in these cells. Comparable experiments in intact animals have also demonstrated that under most physiological conditions HMG-CoA reductase is the key regulated enzyme of cholesterolgenesis *in vivo*.

The regulatory role of HMG-CoA reductase in cholesterol synthesis and the apparent defect in this regulation in familial hypercholesterolaemia (Brown *et al.*, 1981) are two factors which have directed the attention of the pharmaceutical industry to this enzyme for some years, but lingering reservations over the potentially disastrous consequences of inhibiting the pathway prevented a full-scale assault on the enzyme by medicinal chemists.

These fears were largely put to rest in 1976 when scientists at Sankyo announced the discovery of ML-236B (*5.40*), a potent inhibitor of HMG-CoA reductase, which resulted from screening natural products for inhibitors of steroid biosynthesis (Endo *et al.*, 1976). The dihydroxy acid derivative of ML-236B (*5.42*) is a selective, competitive inhibitor of the enzyme with a K_i of 1 nM. Studies at Sankyo and in other laboratories

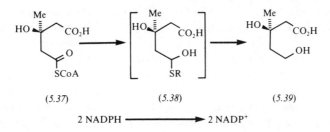

| (5.37) | (5.38) | (5.39) |

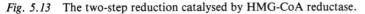

Fig. 5.13 The two-step reduction catalysed by HMG-CoA reductase.

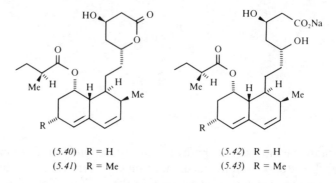

(5.40)   R = H
(5.41)   R = Me

(5.42)   R = H
(5.43)   R = Me

demonstrated profound effects on cholesterol synthesis and, more importantly, on serum cholesterol levels in several animal species. No overt toxicity
was noted (Endo, 1981). This discovery markedly enhanced the level of
interest in inhibition of HMG-CoA reductase and paved the way for the
discovery of other inhibitors.

At least six closely related natural products have now been reported to be
inhibitors of HMG-CoA reductase. The most significant of these is the
methyl derivative (5.41) which was reported almost simultaneously by
scientists at Merck who called it mevinolin (Alberts et al., 1980) and at
Tokyo Noko University who named the compound monacolin K (Endo,
1979). The dihydroxy acid derivative (5.43) of mevinolin is the most potent
HMG-CoA reductase inhibitor yet reported ($K_i = 0.6$ nM).

The structural similarity of the dihydroxy acid portions of ML-236B and
mevinolin to HMG-CoA (5.37) and mevalonic acid (5.39) is undoubtedly
responsible for recognition of the natural products by HMG-CoA reductase.
It is especially intriguing to contemplate the relationship to the hypothetical
enzyme-bound intermediate (5.38). The relative potencies of the various
natural products demonstrate an important but as yet undefined role for the
hexahydronaphthalene and its appended functionality in binding.

Beyond the information conveyed by Fig. 5.13, little structural or mechanistic detail has been elucidated for HMG-CoA reductase. Enzymes from
several sources, including rat liver, have been purified to homogeneity
(Kleinsek et al., 1981). The reduction catalysed by the purified enzyme is a
two-step process in which each of the two hydrides is derived from NADPH.
Following the first reduction, $NADP^+$ is believed to be released and a second
mole of NADPH is bound, but the product of the first reduction step,
mevaldic acid, is not released by the mammalian enzyme. Some workers
believe that a cysteine residue at the active site accepts 3-hydroxy-3-
methylglutaric acid from coenzyme A as the first step in the process.

Sparce as it may be, this mechanistic information should serve as a framework on which to base the design of potential inhibitors. The chemists would

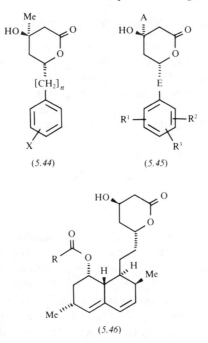

(5.44)

(5.45)

(5.46)

be negligent, however, to ignore the important structural lead represented by ML-236B and its relatives. It is too early to know exactly what approaches have been taken by chemists in the pharmaceutical industry. The scientific and patent literature indicate that early efforts are being directed toward synthetic analogues of the natural products represented by the lactones (5.44), reported by Sankyo chemists (Sato et al., 1980), and (5.45) patented by Merck scientists (Willard et al., 1982). Structural modifications of the natural products themselves have also been reported. For example, various replacements of the 2-methylbutyric ester in mevinolin represented by structure (5.46) have appeared in the patent literature (Willard, 1981).

The utility and safety of HMG-CoA reductase inhibitors for lowering serum cholesterol can only be evaluated in the clinic. Both ML-236B (5.40) (Endo, 1981) and mevinolin (5.41) (Tobert et al., 1982) have been shown to lower serum cholesterol in normal humans. For example, after 7 days' treatment with mevinolin (15 mg, twice daily) serum cholesterol levels fell 25%. If there are toxicity problems associated with interruption of the cholesterol-biosynthetic pathway at the level of HMG-CoA reductase, they are not observed in the acute situation. Whether this inhibition is useful therapy for the prevention of atherosclerosis and heart disease remains to be seen in further clinical trials.

5.2.5   Inhibitors of the renin–angiotensin system

As is evident from earlier sections in this chapter, many important drugs achieve their therapeutic effects through enzyme inhibition. The development of each of these has brought a singular nuance to the term rational drug discovery, but none more so than captopril (5.47), for it is the first clinically available drug to result from the application of an understanding of enzyme mechanisms at the molecular level to drug design.

(5.47)

In the normotensive, blood pressure is controlled by several complex mechanisms. In one such mechanism, the renin–angiotensin system (Fig. 5.14), two proteases, renin and angiotensin-converting enzyme (ACE) are of key importance (Peach, 1977; Laragh, 1978). The sequential action of these enzymes converts the circulating plasma globulin angiotensinogen (5.48), to the octapeptide angiotensin II (5.50), via the decapeptide angiotensin I (5.49). Angiotensin I is biologically inactive, in terms of a direct effect on blood pressure, but angiotensin II is the most potent endogenous pressor substance known. There are at least two mechanisms by which angiotensin II elevates blood pressure: firstly, it has a direct constricting effect on blood

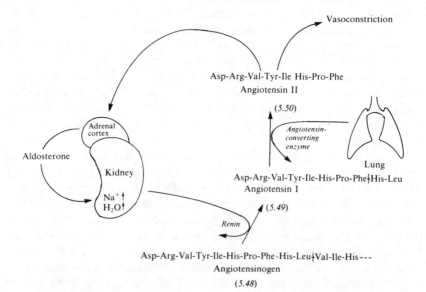

Fig. 5.14   The renin–angiotensin system.

vessels, and secondly, it stimulates the adrenal cortex to release aldosterone which acts on the kidney and results in increased sodium and fluid retention.

The term hypertension refers to a single parameter, blood pressure, but it is a collection of diseases of diverse aetiolgies, rather than a single entity. It is unlikely, therefore, that a single, mechanism-based therapy will be efficacious in all patients. Scientists at the Squibb Institute for Medical Research argued that a defect in the renin—angiotensin system could underlie the disease in at least a subset of hypertensives, those with high renin levels, and that an inhibitor of ACE would lower blood pressure in such patients (Ondetti *et al.*, 1977; Cushman *et al.*, 1977). The Squibb researchers were not alone in making this deduction but most of their competitors rejected the idea, believing that the subgroup of patients most likely to benefit would be difficult to identify and too few in number to ensure commercial success. As sometimes and deservedly happens with bold research programmes, a far broader cross-section of patients respond to the therapy than was first envisaged and although it is still early days, it appears that angiotensin-converting enzyme is an excellent target for drug design.

$$< Glu\text{-}Trp\text{-}Pro\text{-}Arg\text{-}Pro\text{-}Gln\text{-}Ile\text{-}Pro\text{-}Pro$$

(5.51)

In the early phase of the programme, several snake venoms were found to inhibit ACE, and one of these, the nonapeptide SQ 20881 (*5.51*), was shown to lower blood pressure in man. SQ 20881 was inactive by the oral route and was not a practical drug for treatment of hypertension, but it demonstrated the validity of the approach. The problem to be faced was one which seemingly is becoming increasingly common in medicinal chemistry − that of finding small-molecule alternatives to relatively large peptides which achieve the desired pharmacological effect but are poorly absorbed, rapidly metabolized, or both.

Little was known about ACE except that it was a zinc-containing exodipeptidase. However, it was reasoned that ACE might be similar to pancreatic carboxypeptidase A, also a zinc-containing protease. In contrast to ACE, caboxypeptidase A is one of the better understood metalloproteases (Quiocho and Lipscomb, 1971). It is an exopeptidase which selectively cleaves an aromatic amino acid from the *C*-terminus of its substrates. The active site contains a cationic binding site for the carboxylate anion, a lipophilic cleft which accommodates the aromatic amino acid, a proton donor which protonates the *N*-terminus of the departing amino acid, and a zinc cation which activates the amide carbonyl to hydrolysis (Fig. 5.15(a)).

After a lengthy spell of non-productive random screening, the much-needed breakthrough was inspired by the report from Byers and Wolfenden (1973)

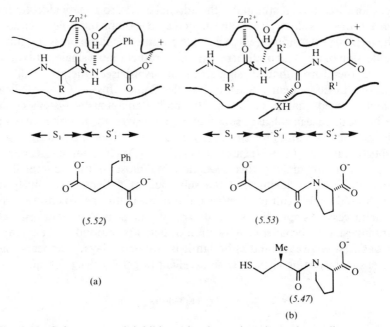

Fig. 5.15 Substrates and inhibitors in the active sites of metalloproteases: (a) Carboxypeptidase A; (b) angiotensin-converting enzyme.

that L-benzylsuccinate (*5.52*) is a potent competitive inhibitor of carboxy-peptidase A. Like Byers and Wolfenden, the Squibb scientists reasoned that the binding of this inhibitor to the active site had much in common with the binding of the substrate and products of the reaction catalysed by carboxy-peptidase A. Since ACE cleaves the *C*-terminal dipeptide from angiotensin I (Fig. 5.15(b)), in contrast to carboxypeptidase A which cleaves a single amino acid from the *C*-terminus of its substrate, they further argued that a succinoyl amino acid, rather than benzylsuccinate, would be a more appro-priate starting point for the design of inhibitors of ACE. Proline was chosen as the terminal amino acid in the target inhibitors because it was the *C*-terminal amino acid in all the naturally occurring peptide inhibitors, e.g. (*5.51*).

The first of the compounds to be synthesized, succinoylproline (*5.53*), was a weak competitive inhibitor of ACE. There followed a number of small structural changes, the most significant of which was replacement of the carboxylate anion in succinoylproline by a thiol group. The drug which quickly emerged from these studies was captopril (*5.47*), which with a K_i of 1.7×10^{-9}M, was not only as potent an inhibitor of ACE as the nonapeptide (*5.51*) first taken to the clinic, but was also orally active.

Captopril was approved by the regulatory authorities in the United States in 1981 and has found a niche in the treatment of hypertension. However, the side effects sometimes associated with its use (blood dyscrasias, loss of taste and skin rashes) have as yet prevented it from assuming a predominant role as an anti-hypertensive agent.

Many other pharmaceutical companies are vigorously pursuing this line of research. A comprehensive review of the structure activity relationships of ACE inhibitors has appeared recently (Petrillo and Ondetti, 1982). At least one 'second generation' inhibitor, enalopril (5.54), a prodrug for the parent acid (5.55), is currently undergoing clinical evaluation (Patchett *et al.*, 1980). Enalopril's inventors have suggested that it is a transition-state analogue, but it appears equally plausible that it too is a bi-product inhibitor, i.e. one which combines some of the binding features present in both products of the enzyme reaction. Interestingly, several of the new inhibitors contain an aryl residue at a position which could bind to the S_1 subsite in the enzyme.

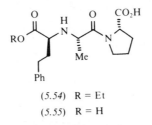

(5.54) R = Et
(5.55) R = H

The future of this exciting new approach to the treatment of hypertension is critically dependent upon the clinical findings with the second generation of ACE inhibitors. Are the side effects produced by captopril unique to that drug, or will they be associated more generally with agents acting by this mechanism? Irrespective of the answer to this question, there is an obvious alternative to ACE inhibitors as a vehicle for attenuating the pressor effects mediated by angiotensin II. Inhibitors of renin should be as effective as ACE inhibitors in dampening synthesis of angiotensin II. Incidentally, studies with renin inhibitors should add to our understanding of the mechanisms by which ACE inhibitors lower blood pressure since ACE catalyses the hydrolysis of bradykinin as well as angiotensin I, whereas renin is without effect on bradykinin. Bradykinin is a potent vasodilator and so ACE inhibitors could affect vessel tone by two mechanisms, reduction in synthesis of angiotensin II, and suppression of the destruction of bradykinin. Current thinking favours the former as the primary mechanism by which captopril lowers blood pressure.

As will be seen, the challenges involved in the search for inhibitors of renin closely parallel those faced by the Squibb scientists a decade ago. Renin is a member of the acid or aspartyl group of proteases (Tang, 1977). Other

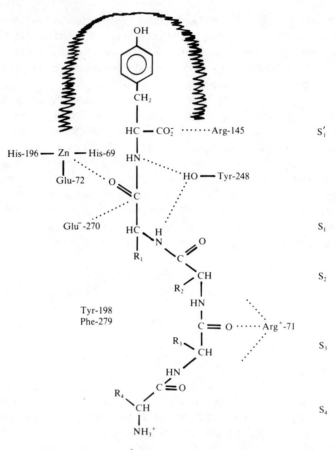

Fig. 5.16 Carboxypeptidase A: detailed interactions within the active site (Quiocho and Lipscomb, 1971; Copyright 1971, Academic Press; used with permission).

members of this class include the mammalian gut enzyme pepsin, and several enzymes of fungal origin, for example from *Pencillium janthinellum* and *Rhizopus chinesis*.

The acid proteases cleave peptides between two hydrophobic amino acids; in the case of human renin the susceptible link is between leucine and valine residues in angiotensinogen. The catalytic mechanism always involves the participation of two aspartyl carboxyl groups at the active site.

There are great practical difficulties in finding inhibitors of renin of therapeutic value. These stem from the differences in primate and rodent substrate and enzyme. Nonetheless, the feasibility of the approach has been clearly demonstrated recently by Cody and colleagues (1980) with the finding that the decapeptide (*5.56*) inhibits renin *in vitro* and prevents the pressor response induced by infusion of human renin in the monkey. The analogy with the Squibb nonapeptide (*5.51*) is obvious.

As was the case with ACE, technical difficulties have prevented detailed structural investigations of renin, and attention has focused on other members of the acid proteases, particularly the fungal enzyme penicillo-pepsin, for which the X-ray structure has been obtained at 0.28 nm (2.80 Å) (Hsu *et al.*, 1977). Again, the contribution made by X-ray studies to our understanding of the fine details of enzyme mechanism has proved to be of fundamental importance (James *et al.*, 1977, 1981).

The mechanisms by which the zinc proteases and the acid proteases hydro-lyse peptides are remarkably similar (Figs 5.16 and 5.17 respectively). In the former the electrophile which activates the amide carbonyl to hydrolysis is the zinc cation, in the latter it is a proton, shared by the carbonyls of Asp-32 and Asp-215 and rendered yet more electrophilic by a charge relay system through the active site to Lys-308. In each case the *N*-terminus of the depart-ing amino acid receives a proton from a tyrosine residue. For penicillopepsin the attacking nucleophile is a bound water molecule and by analogy the same is probably true for the zinc proteases. Both enzymes undergo substantial conformational change on binding substrate.

With the demonstration that the decapeptide (*5.56*) inhibits renin *in vivo*, and the availability of detailed structural information for an acid protease, the analogy with ACE is complete but for an example of a mechanism-based inhibitor equivalent to benzylsuccinate. Pepstatin and analogues provide such examples.

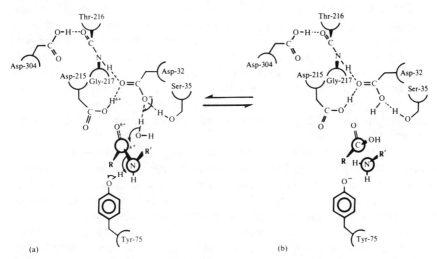

Fig. 5.17 Proposed catalytic mechanism of the acid protease penicillopepsin (James *et al.*, 1977; Copyright 1977, MacMillan Journals Ltd.; used with permission).

Pepstatin (5.57) is a microbial metabolite which inhibits all known acid proteases. The unusual hydroxy acid statine (5.58) occurs twice in the pepstatin molecule. The extremely low inhibition constant observed for pepstatin ($K_i = 4.6 \times 10^{-11}$M, for pepsin) probably derives from the resemblance of statine to the hypothesized transition-state (5.61) for the enzyme-catalysed reaction (Marciniszyn et al., 1976; Marshall, 1976).

$$E + I \underset{k_{-1}}{\overset{k_1}{\rightleftharpoons}} EI \underset{k_{-2}}{\overset{k_2}{\rightleftharpoons}} EI^* \underset{k_{-3}}{\overset{k_3}{\rightleftharpoons}} EI^{**} \qquad (5.1)$$

In an elegant series of investigations Rich and colleagues have probed the mechanism by which pepstatin and analogues inhibit pepsin (1980, 1982). The minimal kinetic scheme which accommodates the facts is given in equation (5.1). Pepstatin (5.57) and some simpler analogues, e.g. the tripeptide (5.59), display a multiphasic pattern of inhibition. Following formation of the collision complex EI, which may be observed by stopped-flow techniques, a second transient EI* forms which is slowly converted to a more tightly bound form EI**. The dissociation constant for the transient complex EI* is approximately 10^{-8}M. Not all pepstatin analogues which bind to the enzyme proceed beyond the transient complex EI* to the tightly bound complex EI**: the ketone analogue (5.60) of the tripeptide (5.59) is such an example.

From an analysis of the inhibition kinetics, the tripeptide (5.59) and pepstatin (5.57) appear to be competitive and non-competitive inhibitors of pepsin, respectively. However, it is more likely that both are competitive inhibitors, the contradictory finding with pepstatin arising from the difficulties in determining the competitive nature of enzyme inhibition for extremely potent inhibitors and in ensuring that the assay system has achieved steady-state conditions. Thus for pepstatin $k_{-3} \simeq$ zero, whereas for the tripeptide (5.59), k_{-3} is sufficiently greater than zero for steady-state conditions to pertain.

Very recently publications have appeared describing X-ray studies at high resolution of pepstatin bound in the active sites of penicillopepsin (Bott et al., 1982) and Rhizopus chinensis carboxyl protease (James et al., 1982). These studies provide a wealth of detail of the binding of this transition-state analogue and add greatly to our understanding of the enzyme mechanism, though they need to be interpreted with caution because of the uncertainty with regards to which of the various possible complexes occurs in the crystalline state.

Of the various pepstatin analogues published to date we consider the

Pro-His-Pro-Phe-His-Phe-Phe-Val-Tyr-Lys

(5.56)

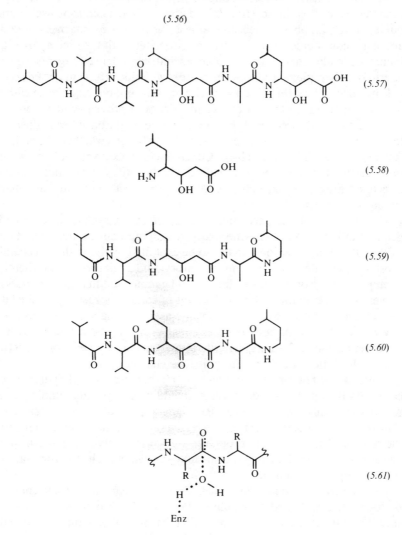

(5.57)

(5.58)

(5.59)

(5.60)

(5.61)

tripeptide (5.59) ($K_i \sim 10^{-9}$M) to contain the most useful structural information in the context of the design of practical mechanism-based inhibitors of renin.

The stage appears set for further exploitation of the renin–angiotensin system as a mechanism-based approach to anti-hypertensive agents. The development of the first inhibitors of angiotensin-converting enzyme will encourage all who choose to base drug design on an understanding of enzyme mechanism.

We have seen from these examples just how important is the choice of target enzyme. It is the keystone of the approach to a clinically useful drug; but it is rarely an obvious decision, at least when new approaches to therapy are being considered. The side effects associated with tyrosine hydroxylase inhibitors make them unacceptable as anti-hypertensive agents because alternative therapies exist, whereas quite severe side effects might be acceptable in a drug where there is no such choice. Without our experience with the non-steroidal anti-inflammatory agents, drugs of proven value before their site of action was discovered, cyclo-oxygenase might have been considered too risky a target for drug action. It remains to be seen whether we shall be as fortunate with inhibitors of HMG-CoA reductase. Concern that only a small proportion of hypertensive patients might benefit, not the potential side effects of drugs, caused many companies to reject the renin−angiotensin system as a target.

Modern techniques, particularly X-ray crystallography, have provided detailed information of enzyme structure and mechanism, but by and large this wealth of information remains to be exploited by the medicinal chemist. Most drugs which are enzyme inhibitors find their origins in chemists' attempts to modify structures thought to be important in cellular processes, or were discovered through some variant of random screening. In the latter case Nature has been of enormous value as a seemingly limitless source of novel structures. Captopril is exceptional in having been tailored to fit a hypothetical model for the active site of one enzyme derived from structural and mechanistic studies on a homologous enzyme.

There can be few better examples than methotrexate and dihydrofolate to exemplify the remarkable differences in binding energies and binding modes which can result from what at first sight appear to be minor structural changes; or in a converse sense, the β-lactam antibiotics, where so many of the structural features once thought essential for activity are absent in the newer agents. The challenge is to be able to predict such structure−activity relationships to advantage.

Almost certainly, small molecule inhibitors of renin and lipoxygenase will be found and evaluated in the clinic. We should welcome good drugs irrespective of their origins, but as medicinal chemists it is natural to hope that they will owe more to rational design than to serendipity.

5.3 RECENT DEVELOPMENTS AND THEIR APPLICATION TO INHIBITOR DESIGN

Medically and economically important drugs which exert their biological effect by modulation of enzyme activity have assured enzyme inhibition an important place in pharmaceutical research. Our understanding of the biochemistry of disease states has increased dramatically over the last decade so that the usual case, as we have seen with 5-lipoxygenase and renin, is that the potentially important enzyme has been identified before the medicinal chemist begins to search for the drug. This has increased the importance of what we have called *ab initio* enzyme inhibitor design. Fundamental advances in our understanding of both the energetics of molecular interactions and the detailed mechanisms of enzyme-catalysed reactions will be of great importance to current and future efforts to design enzyme inhibitors.

If we ignore the more complicated issues of absorption, transport, metabolism and excretion, most problems in drug design reduce to problems of small molecule–large molecule interactions. The lock and key concept of an enzyme and its substrate is no longer a satisfactory model by which to evaluate molecular interactions. Important tools, such as nuclear magnetic resonance spectroscopy, protein X-ray crystallography, interactive computer graphics and quantum chemical calculations, are proving extremely valuable to the medicinal chemist in his efforts to measure these interactions and to harness molecular recognition for enzyme inhibition.

The medicinal chemist will seldom be so fortunate as to have an X-ray crystal structure of the particular enzyme he wishes to inhibit. However, the increasing number of structures of enzymes and enzyme–inhibitor complexes offer many general insights into interactions of small substrates or inhibitors with large proteins (Blundell and Johnson, 1976). By understanding the details of these interactions in a few well-studied examples, the chemist can better formulate an approach to the design of inhibitors of other enzymes.

One of the problems in the past has been the lack of efficient methods for visualizing, manipulating and quantifying the information available from X-ray crystal structures of enzymes. Construction of molecular models of enzymes is a painstaking, time-consuming, expensive undertaking. Developments in computer graphics over the last few years have effectively solved this problem (Langridge *et al.*, 1981; Meyer, 1980). The medicinal chemist can now readily construct computer graphic models of enzymes from atomic co-ordinates. These models can be manipulated in space, unimportant areas of protein structures can be carved away at will and colour-coded three-dimensional active sites can be probed with potential inhibitors. The structures can be viewed as space-filling models to represent steric interactions most accurately or as Dreiding-type models to detail specific molecular interactions most effectively. Interatomic distances and steric congestion, as well as other parameters, can be simply quantified.

Of course, computer modelling systems can also be used effectively for construction of hypothetical enzyme active sites such as the cyclo-oxygenase model of Gund and Shen described earlier. Iterative modification of such models as biological data on each new inhibitor or substrate becomes available should lead to an active-site model as valuable if not as detailed as the actual structure.

Not only do these models provide the medicinal chemist with an efficient means of visualizing enzyme—inhibitor complexes, they also serve as a basis for a much more detailed examination of molecular interactions. Quantum chemical calculations, even at the *ab initio* level, are now possible on fairly large systems (Pincus and Scheraga, 1981; Sheridan and Allen, 1981). These calculations must be used with care since they are still subject to major limitations, for example, the inability to describe solvation effects accurately. They do permit, however, quantification of interactions such as dispersive forces, polarizability and induced dipolar attractions which the chemist finds difficult to handle intuitively. Using point charge libraries for protein residues, electrostatic potential maps can be generated. Complementarity between the electrostatic potentials of the macromolecule and the inhibitor can be examined and probed by structure modification. Energy costs of inhibitor distortion in order to maximize favourable interactions with the enzyme can be evaluated.

One of the major advantages of these modelling systems is that they permit the chemist to evaluate selected parameters of potential inhibitors without actually synthesizing each and every compound. A matrix of many inhibitors can be examined for those which most effectively address particular aspects of design such as steric size and shape, dipole moment alignment, acidity or basicity. If used with caution, modelling, coupled with the medicinal chemist's synthetic expertise, can afford highly efficient enzyme active-site exploration which should expedite the discovery of potent inhibitors of optimal design.

Further insights into enzyme—small molecule interactions are coming from investigations and new interpretations of the kinetics and thermodynamics of enzyme catalysis as was discussed from another point of view by Kluger in Chapter 2. The flavour of exciting developments in this area can be extracted by two examples of current thinking. Albery and Knowles (1977) contemplated the evolution of enzymes over the last 10^9 years. They suggest that the major driving force in an enzyme's evolution is efficiency. It is not appropriate to indulge here in a detailed discussion of the subtleties of Knowles' and Albery's arguments. Basically, they propose that evolutionary single-point mutations could affect enzyme efficiency in three ways. Changes in enzyme structure distant from the active site can *uniformly* affect binding of all the internal states (the enzyme—substrate complex, the enzyme—product complex and transition states). The set of internal states must be adjusted in energy with respect to the unbound product and substrate states (external states). Efficiency is adversely affected by internal states being bound too tightly as well as too loosely. In order to optimize efficiency, enzymes may well have developed some destabilizing interactions with bound states in order to adjust their energies optimally. Recognition of these unfavourable interactions could be used to advantage in inhibitor design.

Additional improvements in efficiency can be obtained by *differential* adjustments to the thermodynamics of the internal states with respect to each other in order to increase flux through the system. For a given enzyme, substrate, intermediate or product states may be specifically stabilized by interactions which developed through evolutionary pressure.

Finally, the only other possibility for increases in efficiency lies in improving catalysis of *elementary steps* by lowering the free energy of kinetically significant transition states. This could be accomplished either by stabilizing transition states or by destabilizing the intermediates leading to them. Knowles and Albery suggest that the argument as to which is the more likely mechanism is irrelevant since, after improvement of a particular catalytic step, changes in uniform binding will optimize the energies of all of the internal states with respect to the external states.

Somewhat different conclusions are drawn by Jencks in his contemplations of enzyme catalysis (Jencks, 1980). He argues that binding of substrates by enzymes must be very tight and exact, giving a complex with exactly the right geometry to react in order to get catalysis, but that much of this enthalpy of binding is used to compensate for the large loss of entropy upon binding. As a result, observed total binding energies may be low, and good substrates may bind even less tightly (as measured by dissociation constants) than poor substrates. An important factor in lowering the activation barrier in the enzyme-catalysed reaction is destabilization of the substrate. This destabilization does not have to be translated into stabilization of the transition state, but the ground-state destabilization must be relieved in the transition state. This idea, and the possibility that these destabilizing interactions can be distant from the reacting centre, considerably broadens the concept of a transition-state inhibitor.

There is great controversy over how enzymes perform their remarkable feats. We have attempted to summarize only two points of view (and have taken some liberties which the original authors might not endorse). These insights into the molecular interactions between substrates and enzymes, coupled with a growing understanding of the reaction co-ordinate leading from substrate through transition states and intermediates to products, are already beginning to make an impact on approaches to the design of enzyme inhibitors as drugs.

The special affinity of enzymes for highly activated intermediates which is necessary for efficient catalysis has led to design of inhibitors which are analogues of these transition states. This approach to enzyme inhibition has been amply reviewed (Lienhard, 1973; Wolfenden, 1976, 1980). Well over 100 examples of transition-state analogue inhibitors have been documented. These analogues, which mimic the geometry and electronics of the transition state but do not collapse to form products, are often very potent enzyme inhibitors with K_i values that can range as low as 10^{-12}M. We have seen earlier, for example, that pepstatin, a potent inhibitor of acid proteases, may mimic the tetrahedral intermediate in peptide hydrolysis.

Mechanistic information which permits the design of transition-state analogues as enzyme inhibitors can also be used for the design of mechanism-based enzyme inactivators. Also called suicide substrates, these inhibitors take advantage of a process which can be summarized as enticing an enzyme to accept a latent inhibitor as a substrate. During the catalytic steps, a reactive intermediate is created which irreversibly inactivates the enzyme. Inhibition of β-lactamase by clavulanic acid is an important example. Many others appear in several recent reviews (Rando, 1975; Abeles and Maycock, 1976; Walsh, 1982).

This approach offers exciting possibilities for drug design but is still subject to some criticism. Walsh (1982) has described mechanism-based inhibitors as the basis for rational design of drugs with maximal specificity *in vivo*. This is probably true for irreversible enzyme inactivators; there is no evidence, however, that suicide inactivators will show enhanced specificity over potent reversible enzyme inhibitors. Additionally, there are concerns over both the immunological consequences of irreversibly modifying proteins *in vivo* and the special problems posed by such inhibitors on overdosage when the target enzyme is the host's, rather than an invading microbe's, and the rate of enzyme synthesis *de novo* may be slow. An important factor in the success of mechanism-based enzyme inhibition is the partition ratio which measures the number of times a suicide substrate is processed to released product without harm to an enzyme molecule per inactivation event. This ratio could be high if the reactive intermediate had time to diffuse away from the enzyme or if the suicide substrate were converted to an innocuous product more often than to a reactive intermediate. The first case presents a potential toxicity problem, the second a potential potency problem.

Potential transition-state analogues and suicide substrates are receiving much attention from medicinal chemists. Such is the lag phase in drug development, however, that it may be several years before the chemists' efforts are rewarded with a commercial success.

REFERENCES

Abeles, R. H. and Maycock, A. L. (1976) *Acc. Chem. Res.*, **9**, 313.
Alberts, A. W., Chen, J., Kuron, G., Hunt, V., Huff, J., Hoffman, C., Rothrock, J., Lopez, M., Joshua, H., Harris, E., Patchett, A., Monaghan, R., Currie, S., Stapley, E., Albers-Schonberg, G., Hensens, O., Hirschfield, J., Hoogsteen, K., Liesch, J. and Springer, J. (1980) *Proc. Natl. Acad. Sci. U.S.A.*, **77**, 3957.
Albery, W. J. and Knowles, J. R. (1977) *Angew. Chem. Int. Ed. Engl.*, **16**, 285.
Appleton, R. A. and Brown, K. (1979) *Prostaglandins*, **18**, 29.

Armarego, W. L. F., Waring, P. and Williams, J. W. (1980) *J. Chem. Soc. Chem. Commun.*, 334.

Baker, D. J., Beddell, C. R., Champness, J. N., Goodford, P. J., Norrington, F. E. A., Smith, D. R. and Stammers, D. K. (1981) *FEBS Lett.*, **126**, 49.

Blundell, T. L. and Johnson, L. N. (1976) *Protein Crystallography*, Academic Press, New York.

Bondinell, W. E., Chapin, F. W., Girard, G. R., Kaiser, C., Krog, A. J., Pavloff, A. M., Schwartz, M. S., Silvestri, J. S., Vaidya, P. D., Lam., B. L., Wellman, G. R. and Pendleton, R. G. (1980) *J. Med. Chem.*, **23**, 506.

Borgeat, P. and Sirois, P. (1981) *J. Med. Chem.*, **24**, 121.

Bott, R., Subramanian, E. and Davies, D. R. (1982) *Biochemistry*, **21**, 6956.

Brown, A. G. (1981) *J. Antimicrob. Chemother.*, **7**, 15.

Brown, M. S., Kovanen, P. T. and Goldstein, J. L. (1981) *Science*, **212**, 628.

Brown, W. E. and Rodwell, V. W. (1980) *Experientia Suppl.*, **36**, 232.

Burger, A. (1980) in *The Basis of Medicinal Chemistry, Part I*, 4th edn (ed. M. E. Wolff), John Wiley and Sons, New York, pp. 1–54.

Byers, L. D. and Wolfenden, R. (1973) *Biochemistry*, **12**, 2070.

Cocco, L., Groff, T. P., Temple, C., Jr., Montgomery, J. A., London, R. E., Matwiyoff, N. A. and Blakley, R. L. (1981) *Biochemistry*, **20**, 3972.

Cody, R. J., Burton, J., Evin, G., Poulsen, K., Herd, J. A. and Haber, E. (1980) *Biochem. Biophys. Res. Commun.*, **97**, 230.

Cushman, D. W., Cheung, H. S., Sabo, E. F. and Ondetti, M. A. (1977) *Biochemistry*, **16**, 5484.

Easton, C. J. and Knowles, J. R. (1982) *Biochemistry*, **12(12)** 2857.

Endo, A. (1979) *J. Antibiot.*, **32**, 852.

Endo, A. (1981) *Trends Biochem. Sci.*, **6**, 10.

Endo, A., Kuroda, M. and Tsujita, Y. (1976) *J. Antibiot.*, **29**, 1346.

Fisher, J., Belasco, J. G., Charnas, R. L., Khosla, S. and Knowles, J. R. (1980) *Philos. Trans. R. Soc. London, Ser. B*, **289**, 309.

Fontecilla-Camps, J. C., Bugg, C. E., Temple, C., Jr., Rose, J. D., Montgomery, J. A. and Kisliuk, R. L. (1979) *J. Am. Chem. Soc.*, **101**, 6114.

Frere, J.-M., Ghuysen, J.-M., Degelaen, J., Loffet, A. and Perkins, H. R. (1975) *Nature (London)*, **258**, 168.

Fuller, R. W. (1982) *Annu. Rev. Pharmacol. Toxicol.*, **22**, 31.

Gund, P. and Shen, T. Y. (1977) *J. Med. Chem.*, **20**, 1146.

Hamberg, M. and Hamberg, G. (1980) *Biochem. Biophys. Res. Commun.*, **95**, 1090.

Hitchings, G. H. and Roth, B. (1980) in *Enzyme Inhibitors as Drugs* (ed. M. Sandler), Macmillan, London, pp. 263–267.

Hitchings, G. H. and Smith, S. L. (1980) *Adv. Enzyme Regul.*, **18**, 349.

Hsu, I.-N., Delbaere, L. T. J. and James, M. N. G. (1977) *Nature (London)*, **266**, 140.

James, M. N. G., Hsu, I.-N. and Delbaere, L. T. J. (1977) *Nature (London)*, **267**, 808.

James, M. N. G., Hsu, I.-N., Hofmann, T. and Sielecki, A. R. (1981) in *Structural Studies of Molecular Biological Interest* (eds G. Dodson, J. P. Glusker and D. Sayre), Oxford University Press, London, pp. 350–389.

James, M. N. G., Sielecki, A., Salituro, F., Rich, D. H. and Hofmann, T. (1982) *Proc. Natl. Acad. Sci. U.S.A.*, **79**, 6137.

Jencks, W. P. (1980) in *Molecular Biology, Biochemistry and Biophysics,* Vol. 32: *Chemical Recognition in Biology* (eds F. Chapeville and A.-L. Haenni), Springer-Verlag, Berlin, pp. 3–25.

Kelly, J. A., Moews, P. C., Knox, J. R., Frère, J.-M. and Ghuysen, J.-M. (1982) *Science*, **218**, 479.

Kleinsek, D. A., Dugan, R. E., Baker, T. A. and Porter, J. W. (1981) *Methods Enzymol.*, **71**, 462.

Knowles, J. R. (1982) in *Antibiotics,* Vol. VI: *Modes and Mechanisms of Microbial Growth Inhibition* (ed. F. E. Hahn), Springer Verlag, Berlin (in press).

Kuyper, L. F., Roth, B., Baccanari, D. P., Ferone, R., Beddell, C. R., Champness, J. N., Stammers, D. K., Dann, J. G., Norrington, F. E. A., Baker, D. J. and Goodford, P. J. (1982) *J. Med. Chem.*, **25**, 1120.

Langridge, R., Ferrin, T. E., Kuntz, I. D. and Connolly, M. L. (1981) *Science*, **211**, 661.

Laragh, J. H. (1978) *Prog. Cardiovasc. Dis.*, **21**, 159.

Laughlin, R. C. and Carey, T. F. (1962) *J. Am. Med. Assoc.*, **181**, 339.

Lienhard, G. E. (1973) *Science*, **180**, 149.

Marciniszyn, J., Hartsuck, J. A. and Tang, J. (1976) *J. Biol. Chem.*, **251**, 7088.

Marquet, A., Frere, J.-M., Ghuysen, J.-M. and Loffet, A. (1979) *Biochem. J.*, **177**, 909.

Marshall, G. R. (1976) *Fed. Proc., Fed. Am Soc. Exp. Biol.*, **35**, 2494.

Matthews, D. A., Alden, R. A., Bolin, J. T., Filman, D. J., Freer, S. T., Hamlin, R., Hol, W. G.-J., Kisliuk, R. L., Pastore, E. J., Plante, L. T., Xuong, N.-L. and Kraut, J. (1978) *J. Biol. Chem.*, **253**, 6946.

Meyer, E. F., Jr. (1980) in *Drug Design*, Vol. IX (ed. E. J. Ariens), Academic Press, New York, pp. 267–298.

Ondetti, M. A., Rubin, B. and Cushman, D. W. (1977) *Science*, **196**, 441.

Ozaki, Y., King, R. W. and Carey, P. R. (1981) *Biochemistry*, **20**, 3219.

Patchett, A. A., Harris, E., Tristram, E. W., Wyvratt, M. J., Wu, M. T., Taub, D., Peterson, E. R., Ikeler, T. J., ten Broeke, J., Payne, L. G., Ondeyka, D. L., Thorsett, E. D., Greenlee, W. J., Lohr, N. S., Hoffsommer, R. D., Joshua, H., Ruyle, W. V., Rothrock, J. W., Aster, S. D., Maycock, A. L., Robinson, F. M., Hirschmann, R., Sweet, C. S., Ulm, E. H., Gross, D. M., Vassil, T. C. and Stone, C. A. (1980) *Nature (London)*, **288**, 280.

Peach, M. J. (1977) *Physiol., Rev.*, **57**, 313.

Petrillo, E. W., Jr. and Ondetti, M. A. (1982) *Med. Res. Rev.*, **2**, 1.

Pincus, M. R. and Scheraga, H. A. (1981) *Acc. Chem. Res.*, **14**, 299.

Quiocho, F. A. and Lipscomb, W. N. (1971) *Adv. Protein Chem.*, **25**, 1.

Rando, R. R. (1975) *Acc. Chem. Res.*, **8**, 281.

Rich, D. H. and Sun, E. T. O. (1980) *Biochem. Pharmacol.*, **29**, 2205.

Rich, D. H., Boparai, A. S. and Bernatowicz, M. S. (1982) *Biochem., Biophys. Res. Commun.*, **104**, 1127.

Roberts, G. C. K., Feeney, J., Burgen, A. S. V. and Daluge, S. (1981) *FEBS Lett.*, **131**, 85.

Samuelsson, B., Goldyne, M., Granstrom, E., Hamberg, M., Hammarstrom, S. and Malmsten, C. (1978) *Annu. Rev. Biochem.*, **47**, 997.

Sato, A., Ogiso, A., Noguchi, H., Mitsui, S., Kaneko, I. and Shimada, Y. (1980) *Chem. Pharm. Bull.*, **28**, 1509.

Schroepfer, G. J., Jr. (1981) *Annu. Rev. Biochem.*, **50**, 585.

Schroepfer, G. J., Jr. (1982) *Annu. Rev. Biochem.*, **51**, 555.

Shen, T. Y. (1981) *J. Med. Chem.*, **24**, 1.

Shen, T. Y. and Winter, C. A. (1977) *Adv. Drug Res.*, **12**, 89.

Sheridan, R. P. and Allen, L. C. (1981) *J. Am. Chem. Soc.*, **103**, 1544.

Strominger, J. L. (1970) *Harvey Lect.*, **64**, 179.

Tang, J. (ed.) (1977) *Adv. Exp. Med. Biol.*, **95**, 1–355.

Tipper, D. J. (1979) *Rev. Infect. Dis.*, **1(1)**, 39.

Tobert, J. A., Hitzenberger, G., Kukovetz, W. R., Holmes, I. B. and Jones, K. H. (1982) *Atherosclerosis*, **41**, 61.

Tomasz, A. (1979) *Annu. Rev. Microbiol.*, **33**, 113.

Vane, J. R. (1971) *Nature (London), New Biol.*, **231**, 232.

Volz, K. W., Matthews, D. A., Alden, R. A., Freer, S. T., Hansch, C., Kaufman, B. T. and Kraut, J. (1982) *J. Biol. Chem.*, **257**, 2528.

Walsh, C. (1981) *Enzymatic Reaction Mechanisms*, W. H. Freeman and Co., San Francisco, pp. 154–520.

Walsh, C. (1982) *Tetrahedron*, **38**, 871.

Willard, A. K. (1981) *US Patent*, 4,293,496.

Willard, A. K., Novello, F. C., Hoffman, W. F. and Cragoe, E. J. Jr. (1982) *US Patent*, 4,375,475.

Wolfenden, R. (1976) *Annu. Rev. Biophys. Bioeng.*, **5**, 271.

Wolfenden, R. (1980) in *Molecular Biology, Biochemistry and Biophysics*, Vol. 32: *Chemical Recognition in Biology* (ed. F. Chapeville and A.-L. Haenni), Springer Verlag, Berlin, pp. 43–61.

Yocum, R. R., Amanuma, H., O'Brien, T. A., Waxman, D. J. and Strominger, J. L. (1982) *J. Bacteriol.*, **149**, 1150.

6 | Metal ions in biological systems

Donald H. Brown and W. Ewen Smith

6.1 INTRODUCTION AND GENERAL CHEMICAL PRINCIPLES

The study of the biological role of metal ions has a long history in medicine, in pharmacology and in toxicology but it is only recently that the extent and variety of metal ion involvement has been appreciated. For example, among the transition metals, the elements V, Cr, Mn, Fe, Co, Ni, Cu, Zn and Mo have been shown to be essential to life, the elements Au, Pt, Pd, Ir, Os, Ti and others have either been used in therapy or claimed to be of therapeutic value and most of the transition metals have been studied with a view to defining and preventing toxicity. As a consequence of these findings most readily available metal ions are now being studied in biological systems for one reason or another.

Since inorganic biochemistry is a socially meaningful area with many different problems, an ever increasing number of chemists with a wide range of specialities have been attracted into the field. On the preparative side, the isolation of specific metalloproteins has triggered off research into the preparation and reactivity of many model compounds such as square-planar cobalt complexes (cf. vitamin B_{12}) and iron–sulphur clusters (cf. ferredoxin). The interest of electronic spectroscopists has been stimulated by such problems as the very high absorbance coefficients found in the 'blue' copper proteins. Kineticists have been involved in studying the catalytic activities of many metalloenzymes, and theoretical chemists have been examining the reasons for the specific reactivities of the metal centres. Thus, the very variety of problems presented by *in vivo* chemistry has provided a fruitful climate for academic researchers with a resultant increased input of effort and output of results. However, the chemistry of metal-containing proteins is complex and our understanding of the action of metal ions in protein environments is patchy, with some processes being much better understood than others. In part this is due to the low concentration at which some metals are present and for which the detection and quantification of the metal is a current research problem in its own right and in part it is due to the uncertainty inherent in relating the role *in vivo* to defined activities *in vitro*. A good deal has been written about some selected and purified model systems such as proteins containing haem groups, vitamin B_{12} coenzyme and zinc enzymes such as carboxypeptidase, but the general question of the impact of metal ion chemistry on our understanding of living systems has been considered only to a much more limited extent. In this chapter, we concentrate on a description of a broad range of metal-containing species, particularly with respect to our understanding of the metabolism, transport and pathology of metal ions.

There are two quite different approaches to the application of inorganic chemistry to biology. Firstly it is possible and in some cases essential to consider the way in which the chemistry of the metal ion affects complete mammalian systems, whole organs or intact cells. For example some square-planar platinum(II) complexes are effective anti-cancer agents (Lippard, 1983) but only *cis* complexes with specific ligand labilities have high activities and it is now generally believed that these complexes work by bridging between two nitrogen bases on DNA by substitution of the more labile pair of ligands. This suggestion gains support from studies *in vitro*, which demonstrated that reactions of this type are feasible within the cell.

Secondly, the chemistry of the metal ion held in a well-defined environment in specific proteins has been investigated successfully *in vitro* in many instances. There are numerous examples of studies in which well-characterized enzymes have been restudied in the environment in which they must function *in vivo*. Both approaches are discussed here with reference to individual elements, two of the elements which have been more exhaustively researched, namely iron and copper, providing most of the examples of studies of individual enzymes.

One aspect of the behaviour *in vivo* of metals which is often ignored by non-chemists is that their chemistry is essentially that of the complexed ion, irrespective of whether more polar ions such as Na^+ or K^+ or more covalent species such as Au(III) or Pt(II) are being considered. Thus, properties such as the effective size and solubility of a metal ion *in vivo* are a function of ligand and solvent present as well as of the metal ions themselves. Further, the correct metal ion balance in various *in vivo* compartments is important for the functioning of specific metal-containing sites in many enzymes and proteins. For example, if the concentrations of some metal ions are considerably raised above the norm, blocking of transport sites can occur and symptoms usually attributed to depletion of other metal ions can appear. Moreover, metal ion distribution can be affected by alterations in the *in vivo* concentration of naturally occurring, low-molecular-weight ligands or of complexing sites in proteins. An example of where this may be the case is in rheumatoid arthritis, where serum histidine levels and albumin thiol levels are significantly lower than in normal subjects. However, as yet little is known about the effect this has on the disease process.

In biological fluids, there are large numbers of complexing species for metal ions. For example, in serum there are many more complexing sites for iron or copper in amino acids, low-molecular-weight peptides and proteins than there are metal ions so that the chemistry of these ions *in vivo* is that of ions present in an excess of competing complexing groups. Thus, considerable impetus has been given recently to the study of complexation in determining likely distributions and reactivities of transition metal ions in biological fluids.

A major problem which emerged in considering *in vivo* activities is that even if the most likely complex species in terms of concentration can be predicted and the relative abundance of these species can be determined, this information may or may not relate directly to physiological activity. There is a wide range of microenvironments present *in vivo* and each would be expected to affect the chemistry of the metal in much the same way as would a specific solvent or the absorption of the metal ions on a particular surface. For example, when transition metals such as copper react with amino acids, quite different complexes are formed in ethanol from those which are formed in water, both because of changes in the solubility of the complexes and because of the different strengths of water and ethanol as ligands. The microenvironments *in vivo* are often separated by membrane barriers which can be penetrated by specific complexes rather than by the ion or other complexes of different size, charge, etc. The platinum drugs already referred to are an example of a situation in which a small proportion of the metal ion present *in vivo* is believed to be in the active species. Milligram quantities are administered but, although the action is believed to be on DNA, fewer platinum ions reach the nucleus than there are DNA molecules present. The remaining platinum is complexed to other cellular and extracellular fractions or is rapidly excreted.

These problems have often been viewed in the past as reasons for ignoring inorganic biochemistry, but the increased understanding of inorganic chemistry and the advent of more powerful physical techniques mean that many more problems can be investigated with a reasonable chance of success. Before looking at the individual metal systems, it seems worth while to outline some of the basic principles developed particularly with reference to transition metal ions.

The classification of metal ions into 'hard' and 'soft' acids is based on

Table 6.1 Classification of metal ions

Na^+, K^+	Mg^{2+}, Ca^{2+}	Transition metal ions
Charge carriers	Structure formers triggers	Redox catalysts and Lewis acids
Mobile	Semi-mobile	Static
Oxygen anion binding	Oxygen anion binding	O/N/S ligands
Weak complexes	Slightly stronger complexes	Strong complexes
Very fast exchange	Moderately fast exchange	Slower exchange

their reactions *in vitro* but is useful also in discussing their *in vivo* chemistry. 'Hard' metal ions are small, and are either not easily oxidized or reduced, or have a relatively high positive charge, e.g. Na^+, K^+, Mg^{2+}, Ca^{2+}, Cr^{3+}. 'Soft' metal ions, on the other hand, are large with a low positive charge, e.g. Cu^+, Au^+, Hg^+, Cd^{2+}, Pt^{2+}. Intermediate between those extremes lie the divalent first-row transition-metal ions. In general, 'hard' metal ions favour complexing with oxygen and nitrogen donors and 'soft' metal ions with sulphur and phosphorus donors. This concept is difficult to apply absolutely but it is useful in a relative sense. For example, copper(II) is 'harder' than copper(I). Therefore, the latter will complex more readily with thiol ligands and a redox enzyme such as caeruloplasmin, in which both oxidation states of copper are required to be stable, uses a mixture of sulphur and nitrogen or oxygen donor atoms to complex the copper. The distinction between 'hard' and 'soft' metal ions is reflected in the biochemical properties of the essential metal ions listed in Table 6.1.

By definition, transition metals have partly filled 'd' orbitals. Thus, strictly speaking zinc(II) (d^{10}) is not a transition metal ion but it is convenient, if not strictly accurate, to include it in this section. The properties of transition metal ions which make them particularly suitable for the types of reactions mentioned above are:

(a) they are good Lewis acids forming wide ranges of complexes with nitrogen, oxygen and sulphur donor ligands, and

(b) because of their unfilled 'd' orbitals they have a readily accessible range of oxidation states available for reduction/oxidation reactions.

The latter is perhaps the major difference between the main group and transition metal ions *in vivo*, but the more covalent nature of the bonding of these ions also produces more directional bonding and quite specific variations in chemical properties which are a function of the nature of the complex rather than of the ligand alone.

Most first-row transition-metal ions prefer octahedral or to a lesser extent tetrahedral geometries. However, where there is a single d electron, distorted structures may be preferred. For example in superoxide dismutase, copper(II) which is a d^9 ion is found in a distorted tetragonal site whereas zinc(II), a d^{10} ion, favours a tetrahedral site. Square-planar configurations and other specific geometries determined largely by the nature of the protein environment are quite common with the consequence that variations in size of the metal can critically affect metal ion reactivity. The divalent ions of the first row have a maximum ionic radius at Mn^{2+} [0.08 nm (0.80 Å)] and a minimum at Ni^{2+} [0.068 nm (0.68 Å)]. An increase in oxidation number usually produces a corresponding decrease in size, Mn^{3+} [0.066 nm (0.66 Å)]. Structural variations can affect other properties such as redox potentials, and catalytically active sites in enzymes are often formed by geometries such as the ferredoxin cluster or the planar haem system which would be regarded as less common in *in vitro* chemistry.

The ligands present in these complexes can affect the reactions of the complex by virtue of their electronic structure as well as their steric preferences. For example, the substitution of $[Co^{III}(NH_3)_6]^{3+}$ by halide ligands gives the stability order $Cl^- > Br^- > I^-$, whereas in $[Co^{III}(CN)_6]^{3-}$ the order is $I^- > Br^- > Cl^-$. The reason for this is that the NH_3 ligand is relatively hard compared to CN^-, and the hardness of the complex is affected by the ligands, altering the affinity of the cobalt for hard or soft ligands. A similar argument may be advanced to explain the relatively soft behaviour of zinc in carboxypeptidase.

There are two ways to consider the stability of complexes, namely kinetic and thermodynamic. The kinetic aspect can often be explained in terms of the ligand-field stabilization energies of the initial complexes and the transition states produced during reaction. For example, Co^{3+}, Cr^{3+} and Ni^{2+} react more slowly than the other first-row transition-metal ions. The thermodynamic aspect is related to the alteration in size of the metal ion and the bonding characteristics of their complexes. This results in a sequence of stability constants for the respective complexes formed between one ligand and various metal ions. This series is known as the Irving Williams order and is $Cr^{2+} < Mn^{2+} < Fe^{2+} < Co^{2+} < Ni^{2+} < Cu^{2+} > Zn^{2+}$. This latter effect is of obvious importance, since the difference in the stability constants of, say, copper(II) and manganese(II) complexes with a ligand may be of the order of 10^3 to 10^5, and copper will substitute for manganese if the availability of the ion is not controlled by other ligands.

One advantage that the study of transition-metal ions has over that of the alkali metals is that a wider range of physical techniques is available, owing to the presence in transition-metal ions of unpaired electrons. These techniques include UV–visible spectroscopy, ESR spectroscopy, magnetic measurements, etc. By means of these and other methods such as NMR, considerable progress has been made in elucidating the role of the transition-metal ions in many enzymic reactions.

6.2 THE TRANSITION ELEMENTS: (1) IRON AND COPPER

6.2.1 Iron

Iron is the most abundant transition metal in the body, with approximately 3–4 g present in adult humans. About 60% of it circulates as haemoglobin, 10% as myoglobin, and 1% is distributed between transferrin and other enzymes. The remainder is mainly in the iron-storage proteins ferritin and haemosiderin.

The aqueous chemistry of iron mainly involves two oxidation states – ferrous (iron(II)) and ferric (iron(III)). The former is a weak reducing agent and is readily oxidized by dioxygen in aqueous solution. Iron(III) in aqueous solution will precipitate out as a polymeric hydroxide at biological pH unless strongly complexed. Both iron(II) and iron(III) form a wide range of complexes with oxygen, nitrogen and sulphur donor ligands. Iron(II) is a d^6 ion and iron(III) a d^5 ion. Both can exist as spin-paired or spin-free species. Six co-ordinate sites are the preferred geometry for both oxidation states, although a limited number of four- and five-co-ordinate complexes are known. Some iron(III) complexes with oxygen ligands such as catechols are inert to substitution and, in general, high-spin iron(II) complexes tend to be more labile than their iron(III) counterparts, and in addition iron(III) hydroxide is insoluble above pH 4 in aqueous solution. These changes in solubility and reactivity are utilized *in vivo* in the transfer and storage of iron.

Iron is absorbed into the body mainly from the gut. With excess iron, the path becomes partly blocked causing an accumulation of ferritin. It is thought that since most iron is likely to be absorbed as iron(II), dietary iron, usually iron(III), has to be reduced and dispersed before absorption (Forth and Rummel, 1973). For example, ascorbic acid enhances the uptake of iron apparently by reducing and complexing it. Powerful complexing ligands, such as EDTA, have the opposite effect probably because they compete strongly with the uptake sites for the available iron. Similarly the poor uptake of iron from vegetable sources is probably due to the strong complexes formed by phytic and oxalic acids. On the other hand, iron from animal sources is better absorbed because a large part of it is haem iron which is readily absorbed unchanged. The presence of peptides and amino acids from protein digestion helps in the absorption, presumably through the formation of complexes of suitable stability to enable subsequent metal ion exchange with storage proteins in the gut.

Iron is transported from the gut by transferrin (a γ-globulin). It has been suggested that the uptake of iron by transferrin involves the oxidation of iron(II) to iron(III) by the copper-containing protein caeruloplasmin and this is thought to be the reason why copper deficiency causes anaemia (Mareschal, Rama and Crichton, 1980). Transferrin then delivers the iron to bone marrow, iron stores and other tissues. The further transfer in these organs is again thought to involve the reduction of iron(III) to the more labile iron(II). Transferrin also cycles iron from the catabolism of haem in the reticuloendothelial system to erythroblasts in the bone marrow. Iron exerts a negative feedback effect on the biosynthesis of transferrin so that in iron deficiency more protein circulates (measured as higher total iron-binding capacity) but with a reduced relative and absolute iron content. Thus, in determining the type of anaemia, both total iron and iron-binding capacity are measured. Transferrin is not thought to have any role other than metal ion transport. It is related to ovalbumin and lactoferrin in that all three proteins have similar amino acid sequences and bind two ferric ions strongly.

The main iron-storage protein in the body is ferritin. Such a reservoir is needed both to provide a source for iron for the many molecules that need it and also to limit the concentration of free iron. These requirements explain the presence of ferritin in most organisms. Iron entering a cell often appears to stimulate the synthesis of apoferritin. This protein is in the form of a spherical ball of external diameter *ca* 12.5 nm (125 Å) with an internal cavity of about 8.0 nm (80 Å) diameter. The structure is made from a symmetrical assembly of 24 polypeptide chains arranged in pairs, each pair lying on the face of a rhombic dodecahedron (Banyard, Stammers and Harrison, 1978). At six of the apices there is a small space about 1 nm (10 Å) across which provides a passage for iron transport. The central cavity has a capacity for over 4000 ferric ions which are present as iron(III) oxide/hydroxide/phosphate aggregates of general formula $(FeO.OH)_8(FeO.H_2PO_4)$. These aggregates vary in size and more than one may be present per molecule. Their orientations do not appear to be related to the protein shell. Extended X-ray absorption edge fine structure (EXAFS) measurements suggest that each iron is co-ordinated by six oxygen atoms. Polymeric iron(III) hydroxides form readily at biological pH and if unrestricted would become too large for a protein shell. To prevent this, the apoferritin incorporates ferrous iron which is then oxidized within the protein shell and polymerizes. The release of iron probably involves reduction to iron(II). The reducing ligands must approach the iron(III) core through the same passages as did iron(II) on incorporation. A number of reducing agents such as thioglycollic acid, cysteine, ascorbic acid and reduced riboflavin have been found experimentally to remove iron (Crichton, 1973). Ferritin is also degraded in lysosomes. Here the protein shell is partly digested, leaving a microcrystalline iron(III) residue of similar composition to that found originally in the ferritin centre. This substance is known as haemosiderin and is the second principal storage form of iron found in animals. In iron deficiency, ferritin and haemosiderin levels drop. As the stores of iron disappear, transferrin concentrations and

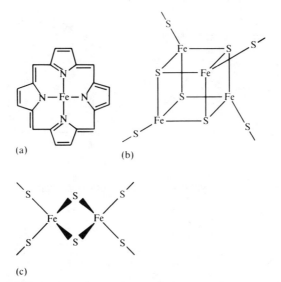

Fig. 6.1   Geometries of some entities containing iron which are found in biological systems. (a) Haem; (b) ferredoxin; (c) rubredoxin.

saturation change. If excess iron is absorbed or if erythrocyte lifetime is reduced, ferritin and haemosiderin appear in many tissues where damage occurs eventually. There does not appear to be any natural method for the removal of excess iron.

The single most important iron co-ordination geometry *in vitro* is that of the haem group (Fig. 6.1) in which the characteristics of the porphyrin ring confer specific properties on the reactive iron at its centre. The unique chemistry which arises has led to many studies attempting to explain both the functions in biology and the physical properties of haem groups (Harrison and Hoare, 1980). For example, the role of haemoglobin circulating in human blood is to provide an efficient uptake of oxygen in the lungs together with an efficient unloading in the body where it is needed. The properties of the molecule are such that, while saturation of the molecule with oxygen in the lungs is not much affected by variations in loading pressure, unloading of oxygen in the tissues is sensitive to local anoxia. This behaviour is related to co-operative interactions between the four subunits of the protein. Haemoglobin contains ferrous iron which reacts with oxygen to form the dioxygen complex. It can be oxidized to the less active ferric form, methaemoglobin. From magnetic measurements it is thought that the deoxy-ferrous complex is a high-spin complex whereas the oxygenated form is a low-spin one, e_g orbitals are slightly larger than t_{2g} orbitals and so the high-spin form of iron(II) is too large to fit into the haem ring, and lies slightly out of plane.

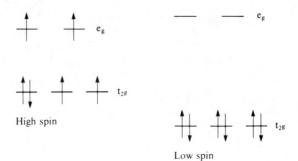

High spin

Low spin

On complexing with oxygen to form a metal–oxygen bond believed to be angled thus:

the iron moves into the ring causing a slight puckering of the ring and a change in protein structure.

Human haemoglobin is tetrameric and the movement of the iron into the plane of the haem ring is transmitted by a series of interactions in the protein to the haem systems in other subunits, increasing their oxygen affinity by favouring the low-spin configuration. Conversely, information from deoxygenated sites will be transmitted and tend to destabilize the low-spin oxygen complexes with respect to the high-spin deoxy complex, thus favouring decomposition of the complex with resultant release of oxygen.

There are many other biologically important haem proteins, usually involved with redox or electron-transport functions. For example, oxygen is reduced by cytochrome oxidase to peroxide within the protein in a manner which protects the environment from attack by reactive intermediates such as superoxide or hydroxide radicals. Cytochrome P-450 and related proteins are involved in hydroxylation reactions.

The iron atoms in haemerythrin, a marine oxygen-transporting pigment, behave somewhat differently. This protein, despite its name, does not contain a haem group and its iron atoms are contained in a single polypeptide chain about 0.35 nm (3.5 Å) apart. The dioxygen-binding affinity of each pair of iron atoms is similar to that in haemoglobin but, in contrast, there is little co-operative action between the different complexing sites. It is thought that dioxygen bridges the two iron atoms in the high-spin iron(II) state to give what can formally be described as two iron(III) atoms with a bridging peroxide group. Both forms can be easily oxidized to a deoxy-genated methaemerythrin protein which contains iron(III) pairs (Klotz, Klippenstein and Hendrickson, 1976).

Another important group of iron-containing proteins are the so-called 'iron-sulphur' proteins (Sweeney and Ranbinowitz, 1980). These generally contain non-haem iron and sulphide ion in equal proportions (Fig. 6.1). They function by electron transfer rather than substrate conversion and examples include the ferredoxins and high-potential iron proteins. They take part in many reactions in bacteria such as nitrogen fixation, ATP formation, photosynthetic electron transfer and also have a role in higher organisms in xanthine oxidase and nitrate and sulphite reduction. The sulphide ions are labile in that they can be displaced as H_2S by acid treatment. This test distinguishes them from the thiol groups also bound to the iron atoms. Three different types of iron–sulphur clusters are commonly found: (a) two iron atoms with two bridging sulphurs; (b) a six-membered ring with alternate iron and sulphur atoms; (c) a cube with alternate iron and sulphur atoms at the corners. In each case the iron–sulphur cluster is linked to the polypeptide chain by cysteine residues with the sulphur being complexed to the iron atoms. The encapsulation of an iron–sulphur group into a polypeptide brings other atoms into close proximity, resulting often in hydrogen-bonding from protein amide groups to sulphide groups. The degree of hydrogen-bonding appears to be significant in controlling the redox potential at which the cluster operates. In each of the iron–sulphur centres there are two accessible redox states, the difference between them corresponding to a one-electron change, thus permitting electron transfer.

Many studies of iron complexes have been stimulated by the very specific *in vivo* chemistries which arise from specific environments. Perhaps the greatest attention has been paid to haem systems with their unusual redox and co-ordination properties, but lately the ability of the ferredoxins to reduce nitrogen, has led to more extensive investigation of cluster systems of this type. These studies have produced novel results and probably form some of the best examples of the way in which biochemistry might be expected to cause an impact on inorganic chemistry as our knowledge of the co-ordination and mode of action of other metal ion species increases.

6.2.2   Copper

Copper is a member of the group 1B elements along with gold and silver. Although in some ways it is a typical transition metal, copper has several characteristic properties. Copper can exist under normal conditions in four oxidation states. Copper(0) (i.e. copper metal) is relatively inert and can be used, for example, in water piping. However, in the presence of oxygen and a strong complexing ligand, copper(0) dissolves easily – an example of this is the ready solubility of copper metal in aqueous solutions of amino acids. This is reputedly the therapeutic mechanism of copper bangles – the dissolution of copper(0) by the amino acids in sweat and the subsequent absorption of the copper–amino acid complexes.

Copper(I) forms mainly 4- and 6-co-ordinate species with essentially tetrahedral or octahedral structure. It is unstable in aqueous solution unless complexed to soft ligands such as thiols or thiourea, but it is stable in some non-aqueous solvents (acetonitrile), and insoluble copper(I) complexes are readily prepared from aqueous solution. As a d^{10} ion, there are no unpaired electrons, with the result that the detection of copper(I) is more difficult than of copper(II) and its role *in vivo* has received much less attention. Nevertheless *in vivo* soft-thiol-complexing sites and non-aqueous environments are suitable for copper(I) and it has been identified in specific proteins such as caeruloplasmin and metallothionein.

The most common oxidation state of copper is copper(II) and being a d^9 ion, its complexes are often distorted resulting in unequal bond lengths and interbond angles and a large variety of possible geometries. Thus, biologically, copper(II) occupies unique sites. Because of the relative stability of its complexes, labile copper concentrations in natural fluids are required to be kept low, otherwise it displaces other transition-metal ions such as Mn^{2+} or Zn^{2+}. Copper(II) also forms dimeric species, the classic example of which is copper acetate (a dimer with bridging carboxylate groups) but dimeric complexes also occur in nature. For example, in haemocyanin, oxygen is carried as a bridged species between two copper(II) ions and not by a haem group as the name would imply.

Copper(III) is relatively rare and is only stabilized by hard ligands such as O^{2-} or F^-, as in $NaCuO_2$ and K_3CuF_6. Whether copper(III) has any role biologically is as yet unknown, but it is possible to form copper(III) *in vitro* in a copper–albumin complex in which the hard ligand is believed to be an N^- atom in the protein chain. The N^- refers to the formal state of the nitrogen in a metal–nitrogen bond.

Copper(I) usually oxidizes rapidly in aqueous solution to copper(II) and thus, copper(I) is most likely to be found complexed to thiols or in lipid-rich tissue or other non-aqueous environments. However, contrary to some textbook statements, copper(I) can be stabilized in aqueous solution with soft ligands such as thiourea, and thiols *in vivo* in combination with the non-aqueous nature of proteins may also enable copper(I) to be present in aqueous compartments. An example of this could be the storage protein metallothionein. Copper(II), on the other hand, is stable in aqueous environments complexed to chloride, water, carboxylate or amino acids. However, since many atoms in enzymes are involved in reduction–oxidation systems, their co-ordination spheres are often composed of a mixture of hard and soft ligands in which both oxidation states are relatively stable. CNDO calculations on thiol—copper complexes suggest that, on going from copper(II) to copper(I), the added electron is distributed over molecular orbitals associated with the ligands as well as with the copper and therefore to speak of formal copper(I) and copper(II) in such complexes is perhaps of little value.

Copper is thought to be absorbed mainly in the gut. Although the chemical pathway is as yet unclear, two routes have been suggested. The first involves diffusion of low-molecular-weight copper complexes and transport by albumin to the liver. The second envisages an important role for the low-molecular-weight, high-sulphur-containing protein metallothionein in the transport of copper from the gut, a mechanism which is currently in favour because it is known that high-zinc(II) dietary content can block copper absorption, and zinc(II) is thought to be transported via metallothionein. The ability of zinc to block copper transport has been used clinically to control copper absorption. The antagonism of molybdenum (particularly thiomolybdates) for copper absorption in ruminant animals is less easy to understand (Bremner and Mills, 1981). Again, metallothionein could be involved. Thiomolybdates(VI) are easily reduced to a dimeric sulphur-bridging molybdenum(V) species which is known to form strong complexes with cysteine — the main amino acid in metallothionein. However, present work on the dietary absorption of copper is not sufficiently definitive to determine whether species of this type are involved.

Copper proteins are involved in a wide range of metabolic pathways and some examples are listed in Table 6.2. The functions given in the table can perhaps be described as indicative rather than comprehensive because many of the enzymes appear to be multifunctional. For example, metallothionein is also a storage protein for zinc and perhaps also a protective protein against the poisonous effect of some heavy-metal ions such as cadmium(II). Caeruloplasmin, which contains up to six copper atoms per molecule, has

Table 6.2 Copper proteins

Storage, transport	Metallothionein
	Caeruloplasmin
Dioxygen transport	Haemocyanin
Electron transfer	Azurin
	Plastocyanics
	Stellacyanin
Mono-oxygenation	Dopamine β-mono-oxygenase
Dioxygenation	Indole 2,3-dioxygenase
Superoxide dismutation	Superoxide dismutase
Substrate oxidation with O_2/H_2O_2	Amine oxidase

variously been described as a storage protein, a peroxidase in iron mobiliz-ation, a transport protein, a serum anti-oxidant (Frieden, 1980). Whether one of these functions is of primary importance and the rest incidental is not known. A slightly different point arises in discussions on superoxide dismutase (Michelson, McCord and Fridovich, 1977) which efficiently catalyses the reaction.

$$2O_2^- + 2H^+ \rightarrow O_2 + H_2O_2$$

It seems strange that this enzyme is required in erythrocytes at a concentration of $3\,\mu M$ in addition to the much higher concentration of glutathione ($2000\,\mu M$) – another efficient free-radical scavenger. Thus, the general role of superoxide dismutase in protecting the cells against free-radical attack must be questioned unless it is located to protect a specific cellular site, as yet unknown, or to generate peroxide at that specific site.

Although the specific functions of the enzymes are still open to question, the role of copper in most of them seems to be concerned with redox reac-tions. Recently the determination of the structure of a number of these enzymes has been of sufficient precision to reveal the immediate copper environment. This has produced a greater understanding of the redox functions of copper in different protein environments. For example, the structure of plastocyanin shows that the copper is present in a flattened tetrahedron which can be considered as intermediate between common geometries of copper(I) and copper(II) (Colman *et al.*, 1977). The four ligands are two histidines complexed through imidazole nitrogens and methionine and cysteine complexed through their sulphurs. Stellacyanin is thought to have a similar structure but with the two sulphurs both arising from cysteines. Thus, slight changes in structure and subtle changes in complexing ligands confer different redox properties on enzymes. The copper sites are close to the surface of the enzymes but not on the surface, producing steric limitations on any further complexation of the copper. Separate anion- and cation-complexing sites have been identified close to the copper, whose role seems to be essentially one of electron transport.

Another type of mechanism is shown by haemocyanins – a group of proteins used to carry oxygen in the blood of the crabs and some other crustaceans. The structures of these compounds are varied depending on source and their functions are complex. However, in each case the copper seems to function as a dioxygen carrier with the ratio of O_2/Cu of $1:2$. In the iron-containing oxygen-transporting enzyme haemerythrin, the O_2/Fe ratio is also $1:2$ as was described earlier. The reason for the apparent duplication of roles of iron and copper in oxygen transport is as yet unexplained.

The complexity of copper reactions and environments is perhaps best illustrated by the copper in the multifunctional enzyme caeruloplasmin (Frieden, 1980). This enzyme accounts for over 95% of the circulating serum

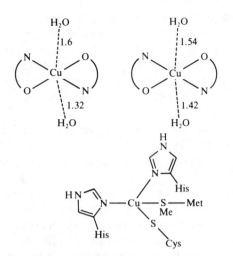

Fig. 6.2 The geometries of copper(II) in complexes can be quite varied. Illustrated are the different bond lengths and angles exhibited by water molecules in *cis* and *trans* bis alanine copper(II) complexes the site in copper blue proteins such as plastocyanin.

copper and since its concentration fluctuates in a number of diseases, it has stimulated considerable research work, both chemical and medical. Its structure has not been completely elucidated, but some idea of the environments of the six copper atoms has been derived mainly from ESR and electronic spectra. The copper atoms have been classified into four different groups. Type I is a copper(II) ion which is assocated with a detectable signal and an intense blue colour; its environment is possibly similar to that described above for plastocyanin (Fig. 6.2). There are thought to be two of these present in each caeruloplasmin molecule. Type II, also copper(II), is ESR-detectable but is not associated with the strong blue colour; it appears to be able to complex anions readily. Type III copper is again thought to be copper(II) but, since it is not ESR-detectable, it is probably in a coupled pair with bridging ligands. Type IV copper is required to make up the total copper content but little is known about it. Since the number of copper atoms in caeruloplasmin varies, this last copper could be associated with copper transport. There is evidence that the copper content of the caeruloplasmin molecule falls in some disease states such as acute rheumatoid arthritis, but the activity of the molecule is unaltered.

The impact of biology on copper chemistry has largely been to stimulate more specific studies of areas where adequate knowledge seemed to exist until the more specific questions related to particular types of biological activity were posed. For example, the geometry of copper(II) complexes is expected to be distorted from octahedral, but the question of the manner in which the different geometries of amino acids would affect this distortion is less clear. One good example of the empirical classification of this effect are amino acid complexes of type $CuL_2(H_2O)_2$ for which a detailed structural analysis of a wide range of compounds is available (Freeman, 1967). The ligand has a considerable effect on the position of the water molecules, and each complex studied is different, but broadly they fall into two types as illustrated (Fig. 6.2).

As in the case of iron, the increasing understanding of the biological role of the metal has led to a large increase in the number of studies of model complexes designed to mimic some aspect of the role of copper *in vivo*. For example, copper Schiff base complexes are believed to be key intermediates in pyridoxal-dependent enzyme processes and possibly in the cross-linking of collagen by lysyl oxidase (Harris, DiSilvestro and Balthrop, 1982). The use of sulphur-containing amino acids produces a more complex series of compounds than that found with alanine etc. and suggests that, as is the case with most of these systems, much remains to be done before *in vitro* knowledge is adequate to begin an interpretation of *in vivo* behaviour (MacDonald, Brown and Smith, 1982). The discovery of the nature of the site in blue copper proteins such as stellocyanin has led to studies of model systems based on this group and the role of copper complexes as oxygen carriers has also been studied. Further, the catalytic role of the copper site can sometimes be effectively studied *in vitro*. For example, copper amino acid complexes function as superoxide dismutases. In medicine, ligands specifically designed to complex and remove copper in Wilsons disease, a condition involving the accummulation of excess copper, have been synthesized and the discovery that copper aspirinate is a more effective and less ulcerogenic anti-inflammatory agent than aspirin has led to the reinvestigation and extension of the chemistry of complexes of this type (Sorenson, 1982).

Another of the limiting factors in such studies is that the chemistry of copper in biological systems is as yet poorly defined. For example, in the discussion of the absorption and distribution of copper, no account was kept of the change in oxidation state of the copper. It seems likely that given the non-aqueous environment and the presence of many thiol groups, copper will often exist as copper(I) and that the body will make use of the different properties of copper(I) and copper(II) in an analogous way to that of iron. However, copper(I) and some forms of copper(II) are difficult to detect and it is clear that a precise understanding of these changes take some time and may require improvements in technique which lie in the province of chemistry rather than biology.

6.3 THE TRANSITION ELEMENTS: (2) OTHER METALS

Some other metals such as zinc and cobalt could equally well be considered in the same detail as iron and copper, but the chemical understanding of the role of metals such as vanadium is still in its infancy. However, enough has been said to establish the relationship between metal ion chemistry and enzyme chemistry and so only the main points relating to the action of other metals will be considered. In the case of elements such as vanadium and chromium, the main emphasis will be on a plausible explanation of the chemical action of these elements in mammalian systems, whereas, for cobalt and zinc, much more specific studies of particular enzymes can be considered.

6.3.1 Vanadium

The interest in vanadium chemistry arose from the experimental observation that ATP supplied by different companies gave different results in terms of the activity of the Na^+/K^+ ATPase (adenosine triphosphatase) required for the action of the sodium pump (Karlish, Beauge and Glynn, 1979). One difference between the samples was the presence in the less-active ATP of vanadium, and further experiments showed that vanadium selectively inhibits Na^+/K^+ ATPase. The vanadium as vanadate seems to be transported into the cells by an anion-transport mechanism. At the inner surface of the cell membrane the vanadate appears to act synergistically with potassium to stimulate binding of the potassium ion. In so doing it alters the response of the sodium/potassium pump to external potassium concentrations so that potassium becomes inhibitory instead of stimulatory.

In spite of this effect, experiments in mice, rats and humans have shown that vanadium complexes are surprisingly non-toxic. However, if large amounts are ingested, as can occur with workers in vanadium mines, widespread inhibition of metabolic processes seems to occur. A possible explanation (Cantley and Aisen, 1979) seems to be that vanadium has two stable oxidation states in normal aqueous solutions. Vanadium(V) exists as the simple vanadate ion $[VO_4]^{3-}$ or as a polymeric form of it, depending on pH. In the presence of weak reducing agents, a complexed vanadyl ion (vanadium(IV)) is easily formed; this contains vanadium bonded to an oxygen giving the ion $[VO]^{2+}$, which then forms five- or six-co-ordinate complexes with a wide range of other ligands. It is thought that there is an NADH-dependent vanadate reduction enzyme present on cell membranes which controls the concentration of vanadium(IV). Thus, alteration of the redox state or the pH of cells could have a greater effect on vanadium metabolism than small variations in dietary intake, but it is also possible that vanadium deficiency or excess can result in the circumvention of normal body control mechanisms.

Recently it has been suggested that vanadium has a role in the aetiology of manic-depressive illness. In this disease, whole body changes in water and electrolyte metabolism have been reported, probably because of the increase in Na^+/K^+ ATPase activity observed in patients with the disease. This latter activity has been found to correlate with plasma vanadium; the plasma vanadium concentrations were high in the manic-depressive state and fell to normal on recovery. However, despite these observations, an understanding of the physiological role of vanadium in tissues is still in its infancy.

6.3.2 Chromium

Chromium is an essential element apparently required for normal carbohydrate metabolism; it is closely associated with insulin metabolism. Normal chromium nutrition levels decrease the requirement for insulin, whereas experimental animals on a chromium-deficient diet have impaired growth, decreased longevity, elevated serum cholesterol and related cardiovascular maladies (Anderson, 1981). Many of these symptoms have also been observed in humans thought to be suffering from chromium deficiency and have been removed by chromium supplementation. Most fresh foods are good sources of dietary chromium with one of the highest concentrations being found in brewer's yeast from which a chromium complex has been isolated but not yet identified (Schwartz and Mertz, 1959). This complex has been found to be capable of potentiating insulin action.

The structure of the active chromium complex is thought to involve nicotinic acid and glutathione or its constituent amino acids, glycine, cysteine and glutamic acid. Synthetic insulin-potentiating chromium complexes have been prepared with glutathione but, again, the exact nature of the active species is unknown because the complexes dissociate in aqueous solution. Part of the problem in this work has been to determine accurately the amount of chromium present in biological fluids. Contamination from stainless-steel equipment and technical problems in chromium analysis are only now being overcome. For example, the accepted value for chromium in serum has fallen in the past few years from about one thousand parts per billion (p.p.b.) to below one p.p.b. With such low concentrations, the isolation and characterization of complexes is very difficult and most of the successful work has depended on the monitoring of biological functions using bioassays. For example, chromium-potentiating factors are being sought by using an insulin assay *in vitro* in the presence of different concentrations of chromium.

The application of inorganic chemistry to this problem can at least help to define the likely types of complex which may be found and the reactions which may occur. Under normal conditions with most ligands such as amino acids the stable oxidation state of chromium is III. It is kinetically and thermodynamically stable and in almost all complexes formed the metal is six-co-ordinate. Chromium(III) can be reduced to chromium(II) but it is only when complexed to ligands, such as cyanide, dipyridyl and thiocyanate, that water-soluble and stable complexes can be formed and it is therefore unlikely that, if nicotinic acid and amino acids are complexed, a chromium(III)−(II) redox system is involved. Thus, with ligands such as nicotinic acid and glutathione, six-co-ordinate chromium(III) complexes are the most likely species *in vivo*.

It has been suggested that the role of the chromium complex is as a catalyst in thiol−disulphide interchange between insulin and sulphydryl membrane acceptor sites. *In vitro*, the slow kinetics of chromium(III) substitution reactions would make this unlikely, but it is always possible *in vivo* that a specific energized reaction does occur. However, another type of chromium system which has not received much mention in the biological literature are species containing bidentate oxygen.

There are quite a number of chromium compounds containing this group, e.g. $K_2Cr(O_2)_2$, pyridine $CrO(O_2)_2$, $(NH_3)_3Cr(O_2)_2$ $K_3[Cr(O_2)_2(CN)_3]$. The formal oxidation state of chromium in these complexes is probably indeterminate. Similarly, the [O−O] group cannot be described accurately as a complexed peroxide or superoxide ion. The binding of insulin to its acceptor site could then be through thiol−disulphide oxidation catalysed by a complexed [O−O] species. However, although inorganic chemistry can help to define the nature of the experimental problem, a precise mechanism of chromium action must wait until further experimental information is forthcoming.

6.3.3   Cobalt

Although cobalt occurs in only a limited number of metal-containing proteins, its role in vitamin B_{12} is vital for a wide range of life forms from micro-organisms to man (Wood and Brown, 1972). Cobalt forms stable complexes with nitrogen-donor ligands, both in the di- and tri-valent states, and both are thought to be implicated in the catalytic cation of the vitamin. The cobalt is complexed in a 'corrin' ring system by four equatorial nitrogen atoms. One axial position is occupied by a donor atom of a side chain of the ring system. The other axial position can be occupied by a variety of ligands such as water, cyanide or a carbon chain, depending on the type of vitamin B enzyme. The discovery of a cobalt−carbon bond in this molecule was unexpected at the time and had an immediate impact in stimulating research into the production of organometallic cobalt complexes.

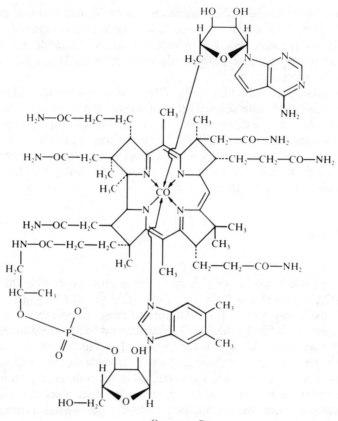

Coenzyme B_{12}

Two of the main catalytic reactions associated with vitamin B_{12} enzymes are hydrogen and methyl transfers. In the former, the mechanism generally involves abstraction of the hydrogen from the substrate by the coenzyme. Substrate rearrangement then takes place and the hydrogen returns to its new position. The cobalt–carbon bond is split prior to reaction with the substrate and is re-formed at the end of the reaction. In the latter group of vitamin B_{12} enzymes, methyl transfer occurs, examples being in the synthesis of acetate and methionine; these use methylated corrin derivatives as intermediates. Methionine synthetase is one example which is found in birds and mammals, as well as bacteria. Its action is to transfer a methyl group from N-methyltetrahydrofolate to homocysteine. The absence or inactivity of this enzyme is thought to contribute to the pathology of pernicious anaemia.

6.3.4 Manganese

The *in vivo* chemistry of manganese seems to be mainly associated with oxygen or one of its reduced forms. Manganese in aqueous solution can exist in a range of oxidation states from (II) to (VII). Of these the most common is the divalent one, and in this state manganese has been used as a probe element in the replacement of zinc(II) from zinc metalloenzymes. In plants, manganese is involved in the respiration and photosynthetic chain. A manganese protein is involved in the abstraction of electrons from water to give dioxygen, and where this enzyme is missing as in certain bacteria, an exogenous reducing agent such as sulphide is required to feed electrons into the chain (Harrison and Hoare, 1980).

Another important manganese enzyme is a superoxide dismutase found in bacteria and often in the mitochondria of more elaborate cells (Michelson, McCord and Fridovich, 1977). The structures of these enzymes are unknown but they contain several manganese atoms with the number depending on where they are found. In common with copper-containing superoxide dis-mutases, their role is to protect tissue from the superoxide ion (O_2^-). However, in doing this they produce peroxide ion (O_2^{2-}), which itself must be removed presumably by an adjacent peroxidase such as glutathione peroxidase or catalase. The manganese superoxide dismutase is not the only body defence against superoxide ion and indeed the copper−zinc superoxide mentioned above has been more widely studied. It is of interest that in Down's syndrome, where the copper−zinc superoxide dismutase concentration is increased, there is a decrease in the manganese dismutase concentration.

6.3.5 Zinc

Zinc exists in solution mainly as a divalent cation. It occurs in a wide range of enzymes and may also be involved in the transfer of information by RNA. It is similar in size to copper(II) and can compete with it in complex formation. However, unlike copper(II) it is not easily reduced. Thus its role in enzymes appears to be as a structure former and, more important, as a Lewis acid in catalytic reactions. Zinc(II), being a d^{10} ion, prefers regular tetrahedral or octahedral geometries. However, in biological systems zinc(II) is sometimes found in a distorted tetrahedral environment necessary for its catalytic role (Bertini, Luchinat and Scozzafa, 1982).

The zinc ion in carboxypeptidase (Dunn, 1975) lies in a shallow hole in the protein which can also accommodate the peptide substrate. The peptide is so oriented that the carboxyl oxygen is positioned in the co-ordination sphere of the zinc(II), which polarizes the carboxyl bond and facilitates nucleophilic attack. Another example of ligand bond polarization occurs in carbonic anhydrase, where the affected ligand is water. Substrate carbon dioxide reacts with the polarized water molecule to give the bicarbonate ion. A similar carbonyl group polarization is important in the chemistry of alcohol dehydrogenase (Chapters 3 and 4).

The role of zinc in superoxide dismutase, on the other hand, appears to be mainly structural. Catalysis is carried out at the copper site, but binding of the zinc(II) ion to the apoprotein makes the enzyme take up a specific structure, resulting in the protein in this region being less accessible to water.

6.3.6 Molybdenum

Molybdenum is unique in that it currently seems to be the only essential second-row transition metal. It occurs in a number of enzymes one of which, nitrogenase, has received a considerable amount of attention because of the obvious economic advantages of low-cost nitrogen fixation (Coughlan, 1980). However, as yet the oxidation states of molybdenum involved in the redox reactions are not fully characterized.

Molybdenum with formal oxidation states from (VI) to (II) can form air-stable complexes. However, although the exact meaning of formal oxidation states is questionable in the presence of Mo=O double bonds because of the delocalization of the electrons, it is likely that the higher states (VI) or (V) will be involved. Molybdenum(VI) exists as the momomers MoO_4^{2-} or in different polymeric forms (involving Mo—Mo bridges) in aqueous solution. In the presence of sulphide, complexes of the type $[MoO_nS_{4-n}]^{2-}$ exist.

Molybdenum(VI) forms complexes in aqueous solution with a wide variety of oxygen donor ligands such as β-hydroxycarboxylic acids and sugars. Molybdenum(V), on the other hand, is more stable with thiol-containing ligands such as cysteine or dithiocarbonate. Many of the molybdenum(V) complexes are dimeric, with bridging oxygen or sulphur atoms. In the presence of cysteine ligands and sulphide ions, molybdenum(V)–cysteine complexes with bridging sulphurs are formed. Stable sulphur donor complexes are also formed with molybdenum(IV) and (III). Molybdenum(II) complexes are generally strong reducing agents, with the odd exception such as the dimeric acetate, $Mo_2(Ac)_4$ which has the same bridging structure as copper(II) acetate.

In nitrogenase it has been assumed without definitive evidence that the molybdenum ion is the site at which the nitrogen molecule is complexed and also reduced directly to ammonia. This nitrogen-fixing enzyme is important in that it provides a major path for the reduction of atmospheric nitrogen to ammonia, thus making nitrogen available to plants and eventually to animals. The enzyme consists of two protein fractions, one containing only iron and the other both iron and molybdenum. Both parts of the protein are very sensitive to oxidation by dioxygen; nitrogen fixing cells which respire aerobically prevent their inactivation by a variety of means generally involving either rapid metabolism of oxygen or the presence of a protein such as haemoglobin, which has a very high affinity for oxygen.

There is still some ambiguity about the nature of the molybdenum present. However, recently EXAFS measurements have shown that the molybdenum is bonded to three or four sulphur atoms at a distance of 0.236 ± 0.002 nm (2.36 ± 0.02 Å) with two or three iron atoms at a distance of 0.272 ± 0.003 nm (2.72 ± 0.03 Å) from the molybdenum (Cramer *et al.*, 1978). There is also evidence for one or two sulphur atoms at a distance of 0.249 nm (2.49 Å). However, further information is still required for an understanding of the mechanics of the electron-transfer systems.

In summary, present knowledge of the chemistry of each transition metal *in vivo* is insufficient to enable a complete picture of their biological activities to be built up. In some cases, such as iron, copper, cobalt or zinc, considerable work has been done, and model studies of specific enzyme sites such as those found for example in vitamin B_{12} or carbonic anhydrase have been extensive and useful. In others, such as vanadium or chromium, much remains to be done before even the nature of the sites involved in the processes *in vivo* can be characterized.

6.4 MAIN-GROUP ELEMENTS, SODIUM, POTASSIUM, MAGNESIUM AND CALCIUM

As mentioned previously the main-group ions are 'hard' in nature and will therefore favour complexing with oxygen anions. They are strongly electropositive and not easily oxidized or reduced. Thus, biologically they have a major role in neutralizing the charge on anions and in maintaining isotonic pressure in natural fluids and are required in relative abundance in living systems. Chemically, Na^+ and K^+ ions are similar, as are Mg^{2+} and Ca^{2+} ions. However, in biological chemistry, small differences in ionic radii are magnified and lead to selective roles for the different cations. For example, K^+ and Mg^{2+} are concentrated in the cytoplasm where as Na^+ and Ca^{2+} are found mainly in serum. Cation-concentration gradients are maintained across cell walls by the expenditure of considerable energy through the hydrolysis of ATP (Racker, 1979). An efficient separation is important in that a rapid influx of Na^+ or Ca^{2+} ions into cells often acts as a trigger for a sequence of biological reactions. The divalent cations Mg^{2+} and Ca^{2+} are also involved in stabilizing certain conformations of enzymes, proteins and membranes. Thus, again, efficient separation is critical. For example, with calcium the intracellular level is 10^{-7} mol dm^{-3} and the extracellular level 10^{-3} mol dm^{-3}. During stimulation, Ca^{2+} acts as a second messenger amplifying the original signal for appropriate cell action. The calcium ion must act at a molecular level by binding to proteins or to enzymes which then modulate the activity of other macromolecules. Two of the more important calcium-binding proteins are calmodulin and troponin C (Wasserman, 1977). The former has a molecular weight of 17 000 and contains four sites for Ca^{2+} ions. The extent of occupancy varies for different reactions. There are also calcium-storage proteins such as calsequestrin, which can bind up to 43 calcium ions per mole of protein. Efflux of Ca^{2+} from these storage proteins can take place in milliseconds to trigger specific reactions. Subsequently, the calcium is pumped back out of the cell by energy supplied by hydrolysis of ATP. Although many calcium-modulated reactions are fairly specific, other divalent cations can interfere. For example, Zn^{2+} inhibits platelet aggregation and phagocytosis, both of which require Ca^{2+} ions. Thus, the overall balance of metal ions *in vivo* is of considerable importance to the proper functioning of the organism. Recently, a range of calmodulin-inhibiting drugs has been developed (Brewer *et al.*, 1982). These apparently function by controlling the influx of calcium ions into cells, with the subsequent inhibition of ATP synthesis and oxygen uptake. Thus, although there are larger concentrations of main-group elements than of transition elements *in vivo*, their environment is often specific and tightly controlled. One example of the synergism between biochemistry and inorganic chemistry which has arisen from these studies has been the continued development of specific complexing ligands such as crown ethers. The main criterion has been to produce ligands with the correct hole size to accept one ion preferentially to another, but more recently the use of complexing groups other than oxygen has been investigated so as to change the chemical affinity of the site rather than its size. As was seen in Chapter 4, crown ethers also have an extensive organic chemistry.

6.5   METAL IONS AS DRUGS: GOLD AND LITHIUM

Many compounds of metal ions have been used as drugs, with compounds of calcium, iron, cobalt or zinc being prime examples. Recently there has been considerable interest in metal ion compounds of non-essential metals, and the use of platinum compounds has already been mentioned. One example of a transition metal (gold) and one of a main-group metal (lithium) are discussed here.

6.5.1   Gold

Gold is a non-essential element but its metabolism is of interest in that several of its complexes find use pharmacologically. The beneficial effects of gold drugs in the treatment of rheumatoid arthritis were confirmed in 1961 by the report of the Empire Rheumatism Council. Since then gold drugs have been quite widely used (Shaw, 1979; Brown and Smith, 1980). There are problems, however, in that some patients do not respond at all and others suffer various toxic side effects. Despite a fairly extensive literature on the clinical aspects of gold therapy, only recently have attempts been made to

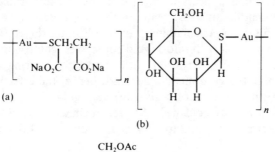

(a)

(b)

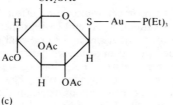

(c)

Fig. 6.3   The structures of the gold drugs myocrisin (a), solganol (b) and auranofin (c). They are used in the treatment of rheumatoid arthritis. The gold is believed to exchange with other thiol ligands *in vivo*. The gold is probably the active species rather than the released thiol.

determine the mechanism of gold action. The gold complexes most commonly used are gold(I)–sodium thiomalate (myocrisin), gold(I)–thioglucose (solganol) and more recently, s-2,3,4,6-tetra-acetyl-1-D-thioglucose (triethylphosphine)–gold(I) (auranofin) (Fig. 6.3). The first two are injected intramuscularly and the latter given orally.

Experiments with animals have shown that after intramuscular injection, gold is widely distributed throughout the body, although higher concentrations are found in organs such as the liver. Gold is retained in identifiable quantities in humans for up to twenty years after gold injections have been stopped. This long-term storage of gold raises the question as to its nature. Experimental evidence has suggested that gold aggregates found in certain cells are probably gold(I)–sulphur species and this seems the most likely type of complex from a comparison with the *in vitro* chemistry of gold.

Gold has two stable oxidation states in aqueous solution, gold(III) and gold(I). The former is stabilized by 'hard' ligands such as chloride or hydroxide and the latter by 'soft' ligands such as thiols, cyanide and thiosulphate. Gold(III) is a fairly strong oxidizing agent and is unlikely to be stable for long in the presence of organic material *in vivo*. Thus, the most likely gold species *in vivo* is gold(I) and the most common ligand the thiol group. Other gold(I)-complexing ligands which occur naturally are cyanide and thiosulphate and there is some evidence that the former may have a part to play *in vivo*.

The role of thiols appears to be of considerable importance in the aetiology of rheumatoid arthritis. Many of the drugs used in its treatment are themselves thiols (D-pencillamine), metabolize to thiols (levamisole or azothioprine) or react with thiols (gold complexes, chloroquine). Thus, the reactivity of different gold drugs with naturally occurring thiols *in vivo* could determine the gold metabolites and, hence, the effectiveness of the drug. Myocrisin and auranofin give different gold distributions in blood, and produce forms of gold with different behaviour *in vivo* such as the ability of the circulating gold to cross cell membranes (Table 6.3).

Table 6.3   The different transporting effect of the gold drugs myocrisin and auranofin. After 12 weeks therapy on either the injected gold drug myocrisin or the oral gold drug auranofin, a sample of blood was taken from each patient and separated into a cellular fraction and plasma. Each fraction was incubated with a placebo fraction of the opposite type. The samples were re-separated and gold concentrations determined by atomic absorption spectrometry. A different fraction of the plasma gold was transferred to the placebo cells indicating different metabolites in the plasma. Gold concentrations are in μg/ml.

Cells	Plasma	Cells	Plasma
Auranofin	Placebo	0.72	0.0
Myocrisin	Placebo	0.28	0.0
Placebo	Auranofin	0.68	0.05
Placebo	Myocrisin	0.05	3.0

Gold(I) forms 1 : 1 complexes with a wide range of thiols including cysteine and glutathione. The structure of these complexes is unknown but is likely to be linear, similar to gold(I)–thiosulphate, with bridging sulphurs. Gold(I) is fairly unique amongst the heavy metals in that most complexes whose structures have been described are linear. This could be due to the small contribution to bonding of the gold 'p' orbitals, leaving the two bonds mainly 's' in character and hence linear. Only when the ligands have more diffuse 'p' orbitals such as with phosphines does greater mixing occur and three- and four-co-ordinate gold(I) compounds are obtained. Gold(III), on the other hand, forms mainly square-planar complexes, as is common for a d^8 metal ion. As mentioned above, gold(III) is a good oxidizing agent and quickly oxidizes thiols to the disulphide, the gold being reduced to gold(I) and stabilized as a thiol complex by excess thiol. If there is no excess of thiol, metallic gold(0) is formed. An interesting exception to this is D-pencillamine (dimethylcysteine). In contrast to cysteine, it forms a stable complex with gold(III). The reason for this is probably steric. Because of the restrictions of the two methyl groups on the carbon atoms adjacent to the sulphur donor atoms, the sulphur atoms will likely be *trans* to each other in the complex whereas with cysteine a *cis* form is possible and any two sulphur atoms *cis* to each other could easily be oxidized to the disulphide. Since many thiols are sterically restricted *in vivo* it is perhaps unwise not to consider the possibility of some gold(III) square-planar thiol complexes occurring. However, such is the affinity of gold(I) for thiols, that the widespread distribution of gold is likely to reflect the widespread distribution of thiols *in vivo*. The other likely ligands mentioned above, *viz.* cyanide and thiosulphate, are present in much lower concentrations in biological fluids. There is some evidence that cyanide in the blood of heavy smokers increases the gold concentration inside cells and hence affects the balance of gold concentrations across cell membranes. However, no work has been published yet on the possible effects of gold on the metabolism of cyanide or thiosulphate or even possibly thiocyanate.

Thus, in conclusion, the therapeutic effect of gold is probably due to a wide range of reactions of which thiol metabolism appears to be the main target. A likely point of action of the gold would be on the structurally important thiol groups on cell membranes.

6.5.2 Lithium

Lithium, the lightest of the alkali metals, is not, as far as is known, an essential element, but lithium carbonate has been found to be effective in the treatment of manic-depressive psychosis (Johnson, 1980). It is thought to function by competing with the naturally occurring metal ions and, because of the different stability of its complexes, alter the existing reaction pathways. The question then arises − with which metal ions will it most likely compete? In aqueous solution lithium will exist as the hydrated Li^+ cation and have the same charge as the Na^+ and K^+ ions. However, it is smaller than these two monovalent cations but is roughly the same size as the Mg^{2+} ion. Using another criterion for comparison, its surface-charge density is similar to that of the Ca^{2+} ion. Thus, depending on the conditions, the Li^+ cation could compete with any of the four naturally occurring cations Na^+, K^+, Mg^{2+} and Ca^{2+}. A complicating factor is that the lithium ion is heavily hydrated in aqueous solution and this affects its complexing ability. For example, with some ion-exchange resins the order of strength of binding has been shown to be $Li^+ < Na^+ < K^+$. Thus, lithium can have a wide range of effects, several of which could have some influence on the same physiological system.

Lithium is the only metal ion which, *in vitro*, is able to replace sodium ion in surrounding fluids in the maintenance, for any significant time, of the resting and action potentials in isolated nerve cells (Birch and Sadler, 1982). The resting potential is mainly due to potassium potential across the nerve cell membrane, and the lithium ion does not interfere significantly with potassium ion transport. However, the action potential is related to the sodium gradient, suggesting that lithium ions may be transported through sodium channels. In general it is thought that lithium ions enter cells in a similar manner to sodium ions but are transported out much more slowly, resulting in intra- and extra-cellular lithium concentrations being of the same order. In the more specific sodium sites, as in the sodium pump, lithium does not compete. Thus, in animal experiments, the addition of lithium ions does not greatly affect serum sodium and potassium ion concentrations.

Lithium ion appears to have an effect on a wide variety of processes, making an interpretation of its biological role somewhat complicated. For example, in rats, serum magnesium and calcium levels increase after lithium injections. Also, lithium often competes directly with calcium and magnesium in the stimulation of enzymes. For example, lithium can replace calcium and stimulate the secretion of acetylcholine.

6.6 MODERN PHYSICAL METHODS

The development of chemistry during the past two decades has produced a battery of powerful techniques which can be applied to problems in inorganic biochemistry. X-ray techniques for example have improved to the point where the environment of a metal in a protein can be elucidated in a much reduced, but still long, time scale. In systems which cannot be crystallized, some information can be obtained by using X-ray absorption edge fine structure (EXAFS). However, other information on such points as rates of reaction, structure in solution and the effect of changing microenvironments is required, and for which the chemist requires a range of techniques which will give a result in the time scale of the experiment. It is only possible to mention a selection of such techniques here, and we have chosen as examples, techniques which we believe will have an increasing impact on inorganic biochemistry.

Modern nuclear magnetic resonance spectrometers are much better equipped for obtaining spectra of direct value in inorganic biochemistry. For example, nucleii such as ^{23}Na and ^{95}Pt can readily be detected, and spin-echo techniques have been used to identify the ^{1}H spectrum of potential ligands such as glutathione in whole cells (Fig. 6.4) (Isab and Rabenstein, 1979). ^{1}H and ^{13}C spectra of metal-ion protein sites in enzymes such as the Cu/Zn superoxide dismutase have been successful in predicting the metal-ion environment in advance of X-ray structural determination, and have the inherent advantage that they do not require single crystals and can be used in principle to determine structure *in situ* or at least in solution in model studies.

Circular dichroism (CD) can be used to define protein configuration in

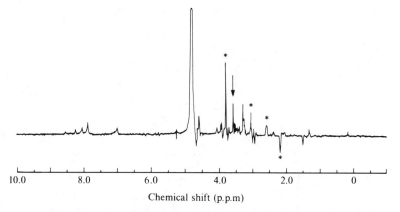

Fig. 6.4 The spin-echo NMR spectrum of intact red cells. This technique utilizes the difference in relaxation times between protons in small molecules and in proteins to produce a relatively simple spectrum of the small molecules. Peaks due to glutathione are marked (*).

such problems as the changes caused in the α and β side chains in haemo-globin in sickle cell anaemia (Woody, 1978). Chiral transition-metal complexes give good (CD) spectra, but good spectra are not restricted to such complexes. The metal ligand bond is partly covalent, and, if there is a chiral centre on the ligand, then there is often sufficient interaction for the for-mally d $\rightarrow$ d transitions on the metal to give intense circular dichroism in the readily accessible visible region. For example, in six-co-ordinate amino acid–chromium(III) complexes, two ligand field bands,

$$^4A_{2g} \rightarrow {}^4T_{2g}(^4F) \text{ and } {}^4A_{2g} \rightarrow {}^4T_{1g}(^4F)$$

appear in the visible-near IR spectrum. In complexes of lower symmetry, such as those with amino acid ligands, these bands are each split into two components which are much more easily identified in circular dichroism than in electronic spectroscopy (Fig. 6.5).

Raman spectroscopy has the advantage in biological systems that measurements of vibrational structure can be taken in aqueous solution. However, it is relatively insensitive, and very sophisticated equipment is required to obtain sufficient signal strength to study most biological systems easily. Extensions of the Raman technique such as coherent anti-Stokes Raman spectroscopy, stimulated Raman or resonance Raman spectroscopy do not suffer from this disadvantage. The most extensively used of these methods at present is resonance Raman spectroscopy in which a laser beam of wavelength similar to that of an absorption band of the molecule is used as the exciting line. An enhancement of intensity of up to about 10^5 com-pared to Raman spectroscopy can be achieved in favourable cases, and since

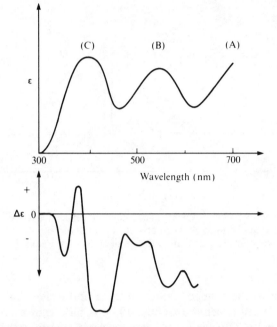

Fig. 6.5 The UV−visible spectrum and circular dichroism of a chromium(III)−serine complex indicating the increased resolution which can be obtained from the latter technique. The UV−visible peaks are due to d → d transitions (A) $^4A_g → ^4T_{2g}$, (B) $^4A_g → ^4T_{1g}(^4F)^4T_{1g}$, (C) $^4A_g → ^4T_{1g}(^4F)$.

the effect is specific for the molecule with the electronic structure which causes the resonance, it is selective.

An example of the use of this technique is the determination of glutathione concentration (about $2000\ \mu M$) in a red cell lysate in the presence of haemoglobin (Banford *et al.*, 1982) (Fig. 6.6). Among the many compounds investigated in this way are several haem-containing proteins, blue-copper proteins and retinal pigments.

Electron-spin resonance can produce good signals with metal ions in biological systems in favourable cases. For example, copper in serum ultrafiltrate ($1-5\%$ of serum copper) and caeruloplasmin can be observed, as can the high-spin paramagnetic forms of haemoglobin (Fig. 6.7) (Peisach *et al.*, 1971).

The main problem with this technique is that the line widths are very sensitive to the metal ion used and its environment. For example in caeruloplasmin, much of the copper is copper(II) and thus should be detected, but it is 'silent', giving no signal as a result of exchange interactions. Thus, although ESR has proved a useful and sensitive probe the results must be interpreted with caution.

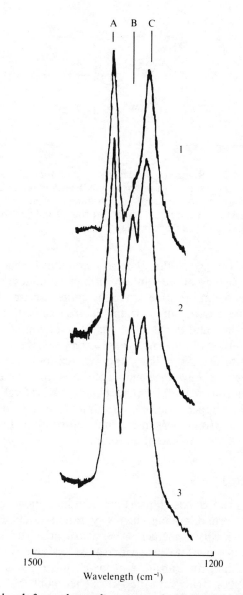

Fig 6.6 Raman signals from a haemolysate treated with 5,5-dithiobis(2-nitrobenzoic acid)(ESSE). The vibrations are produced by haemoglobin and the coloured $ES^{(-)}$ ions produced by reaction of ESSE with intracellular thiol. The electronic spectra of the two overlap but the vibrational spectra shown here are separate and the signal is enhanced so that it can be observed in biological systems at 10^{-5} to 10^{-6}M. A, Haemoglobin; B, ESSE; C, $ES^{(-)}$.

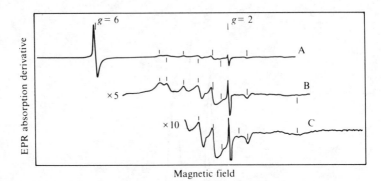

Fig. 6.7 The use of EPR spectra to identify oxidation states and closely related species illustrated in these spectra of haemoglobin in packed red cells. They have been incubated with nitrate to produce a prominent signal due to high-spin ferrihaemoglobin (A) and a variety of low-spin forms (B and C). The total amount of paramagnetic haem is *ca* 5% of haem present.

Finally, one advantage of the study of metal ions is that it is possible to analyse for them in very dilute solutions using techniques that have become rapid, reliable and semi-routine, enabling quite precise pharmacokinetic experiments to be carried out. Perhaps neutron activation analysis is the most reliable method and the ability to use an electron microscope to map metal ion distributions is the most impressive, but the most widely used technique, because of its simplicity and availability, is atomic absorption spectrometry. Fig. 6.8 shows an example of the use of this method applied to the uptake of cuprous oxide by guinea pigs following oral administration. That most of the copper is initially in albumin can be demonstrated by combination of electrophoresis and atomic absorption. Later most of the copper present is bound to caeruloplasmin.

6.7 CONCLUSIONS

Hopefully, this chapter has shown that, although biochemistry was originally an applied form of organic chemistry, recent developments have illustrated that the chemistry of metal ions has also a large part to play and a part which will expand as our knowledge increases. Since many biochemists have little formal teaching in transition-metal chemistry, this has resulted in inorganic chemists moving into the field. Their contributions in the main have been in the elucidation of the environment and reactions of specific metal ions in certain enzymes and the production of an accompanying array of model systems. In a sense this is a logical step for traditional transition-metal complex chemistry, with the same principles probably applying but in a rather more complicated environment.

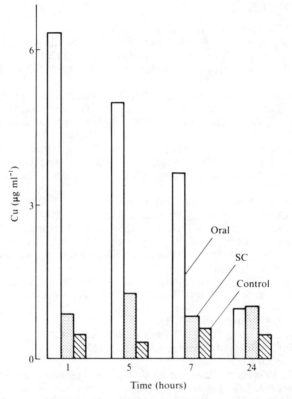

Fig. 6.8 A pharmacokinetic study of the oral absorption of Cu_2O by guinea pigs, illustrating the ease of analysis possible with atomic absorption (SC, subcutaneous injections). The samples were diluted 10 times and injected into the furnace within a few hours of sampling, thus minimizing delays and contamination from sample handling, two of the most important causes of error in this type of analysis.

Running parallel to this have been the investigations into the *in vivo* chemistry, absorption, transport and general metabolism which have been the focus of this chapter. This work involves a more interdisciplinary approach particularly if 'in health' and 'in sickness' variables are also considered. Inevitably this work has lagged far behind the more orthodox chemistry but ultimately should prove the most rewarding. Some impetus for this approach has also come from the now recognized pollution hazards of certain metals and although much more work remains to be done there is the promise that new and exciting insights into biological processes will be forthcoming.

REFERENCES

Anderson, R. A. (1981) *Sci. Total Environ.*, **17**, 13.

Banford, J. C., Brown, D. H., McConnell, A. A., McNeil, C. J., Smith, W. E., Hazelton, R. A. and Sturrock, R. D. (1982) *Analyst*, **107**, 195.

Banyard, S. H., Stammers, D. K. and Harrison, P. M. (1978) *Nature (London)*, **271**, 282.

Bertini, I., Luchinat, C. and Scozzafa, A. (1982) *Struct. Bond.*, **48**, 45.

Birch, N. J. and Sadler, P. J. (1982) in *Specialist Periodical Reports in Inorganic Biochemistry*, Vol. 3 (ed. H. A. O. Hill), Royal Society of Chemists, London, pp. 372–396.

Bremner, I. and Mills, C. F. (1981) *Philos. Trans R. Soc. London, Ser. B*, **294**, 75.

Brewer, G. J., Bereza, U., Kretzchmar, P., Brewer, L. F. and Aster, J. C. (1982) in *Inflammatory Diseases and Copper* (ed. J. R. Sorenson), Humana Press, New Jersey, pp. 532–535.

Brown, D. H. and Smith, W. E. (1980) *Chem. Soc. Rev.*, **9**, 217.

Cantley, L. C. and Aisen, P. (1979) *J. Biol. Chem.*, **254**, 1781.

Colman, P. M., Freeman, H. C., Guss, J. M., Murta, M., Norris, V. A., Ramshaw, J. A. M. and Venkatappa, M. P. (1978) *Nature (London)*, **22**, 319.

Coughlan, M. (ed.) (1980) *Molybdenum and Molybdenum-Containing Enzymes*, Pergamon Press, Oxford.

Cramer, S. P., Gillum, W. O., Hodgson, K. O., Moretensen, L. E., Stiefel, E. I., Chisnel, J. R., Brill, W. J. and Shah, V. K. (1978) *J. Amr. Chem. Soc.*, **100**, 3814.

Crichton, R. R. (1973) *Struct. Bond.*, **17**, 67.

Dunn, M. F. (1975) *Struct. Bond.*, **23**, 61.

Forth, W. and Rummel, W. (1973) *Physiol. Rev.*, **53**, 724.

Freeman, H. C. (1967) *Adv. Protein Chem.*, **22**, 257.

Frieden, E. (1980) *CIBA Found.*, **79**, 93.

Harris, E. D., DiSilvestro, R. A. and Balthrop, J. A. (1982) in *Inflammatory Disease and Copper* (ed. J. R. J. Sorenson), Humana Press, New Jersey, pp. 183–199.

Harrison, P. M. and Hoare, R. J. (1980) *Metals in Biochemistry*, Chapman and Hall, London.

Isab, A. A. and Rabenstein, D. L. (1979) *FEBS Lett.*, **106**, 325.

Johnson, F. N. (ed.) (1980) *Handbook of Lithium Therapy*, University Park Press, Baltimore.

Karlish, S. J. D., Beauge, L. and Glynn, I. M. (1979) *Nature (London)*, **282**, 333.

Klotz, I. M., Klippenstein, G. L. and Hendrickson, W. A. (1976) *Science*, **192**, 335.

Lippard, S. J. (ed.) (1983) *Biochemistry of Platinum, Gold and other Therapeutic Agents* (American Chemical Society symposium, no. 209), American Chemical Society, Washington D.C.

MacDonald, L. G., Brown, D. H. and Smith, W. E. (1982) *Inorg. Chim. Acta*, **67**, 7.

Mareschal, J. C., Rama, R. and Crichton, R. R. (1980) *FEBS Lett.*, **110**, 268.

Michelson, A. M., McCord, J. M. and Fridovich, I. (eds) (1977) *Superoxide and Superoxide Dismutase*, Academic Press, New York.

Peisach, J., Blumberg, W. E., Ogawa, S., Rachmilewitz, E. A. and Oltzik, R. (1971) *J. Biol. Chem.*, **246**, 3342.

Racker, E. (1979) *Acc. Chem. Res.*, **12**, 338.

Schwarz, K. and Mertz, W. (1959) *Anal. Biochem. Biophys.*, **85**, 292.

Shaw, C. F. (1979) *Inorg. Persp. Biol. Med.*, **2**, 287.

Sorenson, J. R. J. (ed.) (1982) *Inflammatory Diseases and Copper*, Humana Press, New Jersey.

Sweeney, W. V. and Ranbinowitz, J. C. (1980) *Annu. Rev. Biochem.*, **49**, 139.

Wasserman, R. H. (1977) *Calcium Binding Proteins and Calcium Function*, Elsevier, Amsterdam.

Wood, J. M. and Brown, D. G. (1972) *Struct. Bond.*, **11**, 47.

Woody, R. (1978) in *Biochemical and Clinical Aspects of Haemoglobin Abnormalities*, Academic Press, New York, pp. 279–298.

7 | Enzyme-level studies of the biosynthesis of natural products

David E. Cane

7.1 INTRODUCTION

It is over ninety years since J. N. Collie (Collie, 1893) first suggested a connection between the reactions of simple carbonyl-containing compounds and the structures of numerous aromatic plant metabolites. In the intervening years biogenetic theory has expanded and matured so that today it is possible to rationalize, at least in principle, the origins of the vast majority of naturally occurring organic substances. The introduction of radioisotopic tracers and the more recent revolution in stable isotope nuclear magnetic resonance methods has provided a firm experimental base for the widely accepted precursor–product relationships between the simple compounds of central metabolism, acetate, amino acids and carbohydrates, and the seemingly endless variety of organic natural products. Over the last few years the study of the biosynthesis of natural products has entered a new and extremely exciting phase. After many years of working almost exclusively with intact cells or whole organisms, an increasing number of investigators have begun to turn their attention to the individual enzymes of secondary metabolic pathways. This development has occurred so rapidly, that it is already impossible even to summarize all the numerous important examples of enzyme-level studies which have been reported recently and which have led to significant advances in our understanding of the origins of natural products. It is nonetheless possible to consider some of the motivations for this work, to inquire into both its potential advantages and disadvantages, and to reflect on the promise it holds for the future development of bio-organic chemistry.

It might well seem odd that a discipline whose major concern has been to uncover the ways in which Nature carries out organic chemistry should have taken so many years to turn its attention to Nature's primary agents, the enzymes themselves. In fact, many biosynthetic problems have historically been investigated at the enzyme level. The classic investigations of steroid biosynthesis carried out in the 1950s and 1960s (Bloch, 1952; Popjak and Cornforth, 1960, 1966; Clayton, 1965; Cornforth, 1973) have led to a detailed understanding of the intricate pathway by which acetyl-CoA is transformed to cholesterol by way of the key intermediates mevalonic acid and squalene. Most of the enzymes of this pathway have been characterized, some in crystalline form and others still only as crude preparations; the results of these landmark studies have served as an experimental and conceptual touchstone for numerous investigations of less-well-defined terpenoid systems. The major outlines of the porphyrin-biosynthetic pathway were defined from the outset at the enzyme level (Bogorad, 1966; Burnham, 1969), while more recent work on the marvellously rich arachidonic acid cascade has relied predominantly on enzymological investigations (Samuelsson et al., 1978; Samuelsson, 1981). On the other hand, the enzymology of polyketide, terpenoid and alkaloid biosynthesis is only in its relative infancy, in spite of the fact that these classes of compounds were not only the focus of early biogenetic speculation but have remained at the traditional centre of biosynthetic investigation.

In each case, the biosynthesis of compounds belonging to these three major classes of metabolites has been studied primarily by feeding isotopically labelled precursors to growing cells or otherwise intact tissues followed by isolation of the resulting labelled metabolites and appropriate analysis, either degradative or spectroscopic, to determine the sites of labelling (Brown, 1972). Incorporation experiments of this type, many of which have achieved the pinnacle of intellectual elegance and experimental finesse, have led not only to the identification of the primary precursors of individual polyketides, terpenoids and alkaloids, but have also allowed plausible deductions as to the probable pathways by which the simple precursors are transformed to the more complex natural product. Guiding the design and interpretation of all these incorporation experiments has been a small number of central hypotheses which form the core of biogenetic theory and which are thought to summarize the major mechanisms of carbon–carbon and carbon–heteroatom bond formation. For example, cyclic polyketides are thought to arise by cyclization of linear β-polycarbonyl chains by way of simple aldol-type condensations followed by appropriate dehydrations and reductions (Fig. 7.1(a)). Similarly, cyclic terpenoids are seen as derivable from a small group of acyclic precursors by a series of intramolecular electrophilic additions to double bonds followed by suitable deprotonations, reprotonations, hydride shifts and Wagner–Meerwein rearrangements

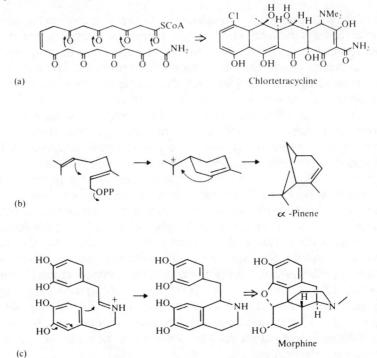

(a)

Chlortetracycline

(b)

α -Pinene

(c)

Morphine

Fig. 7.1 Polyketide, terpenoid and alkaloid cyclization prototypes. (a) Conversion of linear decaketide to chlortetracycline; (b) cyclization of geranyl pyrophosphate to α-pinene; (c) formation of morphine from Schiff's base.

(Fig. 7.1(b)). The characteristic molecular frameworks of the alkaloids may frequently be derived by Mannich-type reactions of an appropriate nucleophilic enolic or aromatic double bond with an iminium ion, itself generated by simple condensation of an amine with a carbonyl group (Fig. 7.1(c)). The enormous success of these central generalizations in predicting as well as rationalizing such a massive body of experimental results, along with the host of biomimetically modelled syntheses inspired by these paradigms, has left little doubt as to the essential correctness of these biogenetic hypotheses. Nonetheless, it is only within the last few years, as the enzymes of secondary metabolism have come under experimental scrutiny, that biosynthetic investigators have begun to confront directly the key biosynthetic events which lie at the heart of biogenetic theory.

The potential advantages of enzyme-level studies will be apparent if one considers briefly the nature of the experimental system with which the biosynthetic investigator is usually forced to work. On the basis of the intact and specific transformation of some precursor, be it acetate, mevalonate or an amino acid, to a distant polyketide, terpenoid or alkaloid product, a pathway may be proposed involving a series of discrete chemical intermediates, each of which is presumed to be the product of one specific enzyme and the substrate for a second. Various structural criteria have evolved for establishing the actual sequence of intermediates along the pathway, including the isolation of the candidate compounds themselves and the demonstration that each substance not only acts as a precursor of the end product of the pathway, but is derived from the common precursor as well. Further kinetic criteria must also be satisfied, such as the requirement that label from the common precursor must reach the intermediate before the end product, and that the putative intermediate must be converted to end product faster than is the primary precursor. The experimental problem is complicated because potential intermediates must be able to traverse various cellular barriers in order either to diffuse out of, or to re-enter, the metabolizing cell. The problem is particularly severe with highly reactive substrates which may not be sufficiently long-lived to allow detection, or with highly polar substances which, devoid of any active transport system, are unable to traverse the normally lipophilic barriers of the cell. The advantages of a cell-free system are obvious and three recent examples will serve to illustrate the power of the method.

7.2   RECENT ADVANCES IN THE USE OF CELL-FREE SYSTEMS

7.2.1   Biosynthesis of vitamin B_{12}

By 1974, extensive ^{13}C NMR investigations in the laboratories of Battersby and Scott had established aminolaevulinic acid (7.1), porphobilinogen (7.2) and methionine (7.3) as the basic building blocks of the corrin ring system of vitamin B_{12} (7.5) and had strongly suggested that uroporphyrinogen (urogen) III (7.4) might be the key link between the porphyrin and corrin families (Fig. 7.2) (Battersby and McDonald, 1975; Scott, 1975). In spite of the reasonableness of this hypothesis, numerous attempts at experimental verification of the presumed role of urogen III by feedings to intact cells had been unsuccessful (see, however, Scott et al., 1972). The development of a viable cell-free preparation from two vitamin B_{12}-producing organisms Propionibacterium shermanii (Scott et al., 1973; Battersby et al., 1975) and Clostridium tetanomorphum (Dauner and Mueller, 1975) not only unambiguously proved that urogen III could be converted to the corrin ring system, but eventually led to the isolation and identification of several key intermediates in this conversion. The first of these new metabolites to be recognized was sirohydrochlorin (Factor II) which was shown to have structure (7.7) (Fig. 7.3) containing methyl groups on rings A and B respectively (Deeg et al., 1977; Scott et al., 1978; Battersby et al., 1978). Subsequently two additional methylated urogens were isolated and characterized. Factor I proved to be a mono-methyl derivative (*7.6*) (Mueller *et al.*, 1979a) while Factor III (*7.8*) was shown to carry three additional methionine-derived methyl groups, one each attached to rings A and B and the third, quite unexpectedly, at C-20 which must be lost in a subsequent corrin-forming ring contraction (Mueller *et al.*, 1979a, b; Battersby *et al.*, 1979; Lewis *et al.*, 1979). The recent demonstration by teams in Zurich (Mombelli, 1981) and Cambridge (Battersby

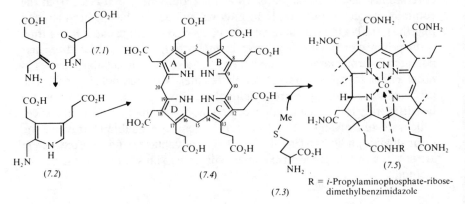

Fig. 7.2 Conversion of aminolaevulinic acid (*7.1*), porphobilinogen (*7.2*) and methionine (*7.3*) to vitamin B_{12} via urogen III (*7.4*).

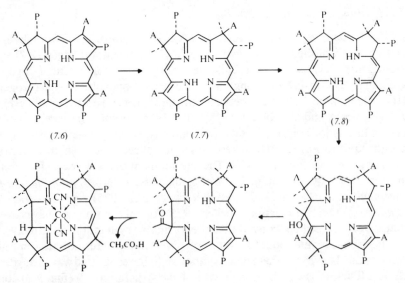

A = CH_2CO_2H, P = $CH_2CH_2CO_2H$

Fig. 7.3 Conversion of Factors I, II and III to cobyrinic acid.

et al., 1981) that C-20 and its newly acquired methyl partner are extruded exclusively as acetic acid has led to a new and highly reasonable proposal for the *mechanism* of the corrin ring contraction based on the classical semi-benzoin arrangement (Fig. 7.3) (Rasetti *et al.*, 1981). Moreover, subsequent work with ever more highly purified enzyme preparations has established the actual oxidation states of Factors I and II during their enzymic transformations as the corresponding tetrahydro and dihydro derivatives respectively (Mueller *et al.*, 1979a; Battersby *et al.*, 1982). Most recently, Battersby has used a combination of high-field NMR and pulse–chase techniques to deduce the fact that the next methylation occurs at C-17 (Uzar and Battersby, 1982). It should be apparent that none of these studies would have been feasible using traditional whole-cell feeding methods.

7.2.2 Biosynthesis of penicillins and cephalosporins

The second example of the potential of cell-free investigations comes from the study of the biosynthesis of the enormously important penicillin and cephalosporin families of β-lactam antibiotics. Once again prior work, first with radioactive and subsequently with [^{13}C]-labelled substrates, had firmly established the amino acids valine, cysteine and α-aminoadipic acid as the common precursors of penicillin N (*7.9*) and cephalosporin C (*7.10*) (Aberhart, 1977; Fig. 7.4). Here too, a likely intermediate had been identified in the form of the Arnstein tripeptide, δ-(L-α-aminoadipyl)-L-cysteinyl-D-valine (*7.11*) [(LLD)-ACV] (Arnstein and Morris, 1960). Nonetheless all attempts to demonstrate intact incorporation of the simple tripeptide into

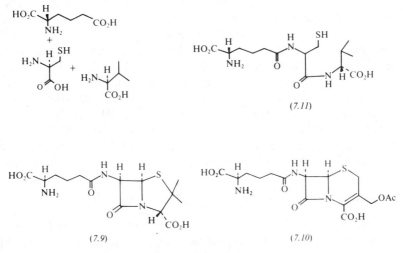

Fig. 7.4 Origin of penicillin N (*7.9*) and cephalosporin C (*7.10*).

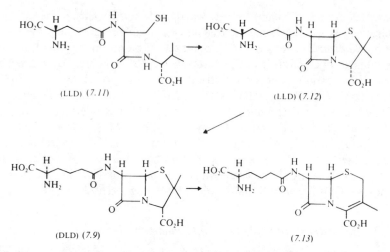

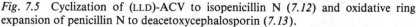

Fig. 7.5 Cyclization of (LLD)-ACV to isopenicillin N (7.12) and oxidative ring expansion of penicillin N to deacetoxycephalosporin (7.13).

penams or cephems using cultures of *Pencillium chrysogenum* or *Cephalosporium acremonium* failed. The problem was eventually overcome, not without considerable effort, when Abraham discovered that viable cell-free preparations of the key cyclase could be obtained by osmotic lysis of protoplasts of *C. acremonium* (Fawcett *et al.*, 1976). A further puzzle in β-lactam biosynthesis was solved when it was found upon careful analysis that the initial product of the cyclization of (LLD)-ACV was isopenicillin N (7.12), containing an unrearranged L-α-aminoadipyl side chain (O'Sullivan *et al.*, 1979; Konomi *et al.*, 1979), rather than the epimeric penicillin N, as had been previously supposed (Fig. 7.5). In spite of the claimed isolation of a monocyclic β-lactam intermediate (Meeschaert, 1980), to date all other attempts to detect further intermediates between the linear tripeptide and the bicyclic penam system, including direct analysis by [^{13}C]-NMR, have been unsuccessful (Baldwin *et al.*, 1980a). In fact it is not yet known whether the ACV cyclase consists of one or more enzymes. Further work (Baldwin *et al.*, 1980b, 1981) has confirmed an earlier claim (Kohsaka and Demain, 1976) that penicillin N is converted by a cell-free preparation from *C. acremonium* to deacetoxycephalosporin C (7.13), thereby establishing the long-suspected direct link between the penicillin and cephalosporin families of antibiotics.

7.2.3   Biosynthesis of indole alkaloids

The third example of the contributions made by the use of cell-free systems comes from the area of indole alkaloid biosynthesis. During the mid-1960s rapid progress was made in establishing the cyclopentanoid monoterpene

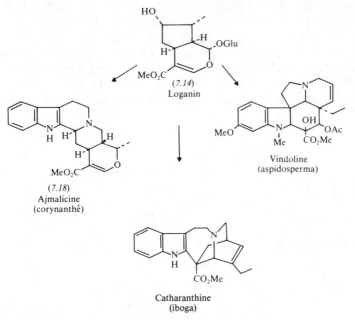

Fig. 7.6 Derivation of indole alkaloids from loganin (*7.14*).

loganin (*7.14*) as the long-sought precursor of the ten-carbon non-amino acid-derived fragment of the corynanthé, aspidosperma and iboga alkaloids (Fig. 7.6) (Battersby *et al.*, 1969, 1968a, b; Brechbueler-Bader, 1968; Loew and Arigoni, 1968), and extensive isolation and feeding experiments with *Catharanthus roseus* and other indole-alkaloid-producing plants had suggested a plausible series of intermediates for the conversion of loganin to the major indole alkaloids (Battersby, 1967, 1971; Scott, 1970). One particularly puzzling problem, however, was the difficulty in explaining how vincoside (*7.16*), the 3β-epimer of the condensation product of tryptamine and secologanin (*7.15*) (Blackstock *et al.*, 1971), rather than its 3α-epimer strictosidine (*7.17*) (isovincoside) (DeSilva *et al.*, 1971), could be converted to the various indole alkaloids without loss of the C-3 hydrogen (Battersby *et al.*, 1968c, 1969; Battersby and Gibson, 1971), in spite of the requirement for an epimerization at this centre. With the isolation of a cell-free enzyme system from tissue cultures of *C. roseus* capable of mediating the formation of ajmalicine (*7.18*) and related corynanthé alkaloids (Scott and Lee, 1975; Stoeckigt *et al.*, 1976) came the surprising but gratifying finding that the earlier precursor relationships had in fact been somehow reversed and that strictosidine was the actual biological precursor (Stoeckigt and Zenk, 1977; Scott *et al.*, 1977) (Fig. 7.7). Not only has strictosidine synthetase now been purified some 740-fold to apparent homogeneity (Mizukami *et al.*, 1979), but the sequence of intermediates leading to ajmalicine has now been firmly established (Fig. 7.8). Notably, the role of geissoschizine (*7.20*), long a matter of controversy, has been clarified by cell-free experiments which show that it is an apparent shunt metabolite in redox equilibrium with dehydrogeissoschizine (*7.19*) which lies along the main pathway (Stoeckigt, 1978; Stoeckigt *et al.*, 1980; Rueffer *et al.*, 1979). These conclusions have in fact been supported by isotope competition experiments which establish that

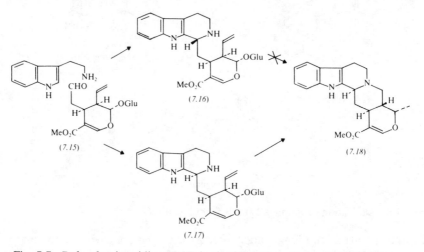

Fig. 7.7 Role of strictosidine (*7.17*) in the formation of ajmalicine (*7.18*).

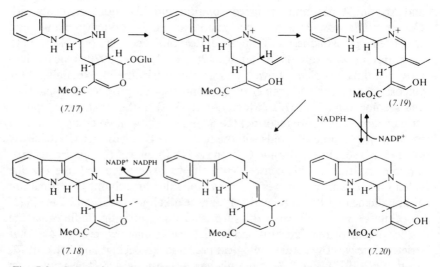

Fig. 7.8 Conversion of strictosidine (*7.17*) to ajmalicine (*7.18*) via dehydrogeisso-schizine (*7.19*).

ajmalicine is labelled more rapidly by tryptamine than by geissoschizine (Lee *et al.*, 1979). The latter experiments avoided the ambiguities in uptake and transport inherent in whole-cell competition experiments and emphasize the power of cell-free techniques in clarifying subtle questions of metabolite sequence, traditionally one of the most difficult of all biosynthetic problems to resolve. It will also be noted that only work at the cell-free level can directly establish the cofactor requirements of biosynthetic transformations, a knowledge of which is essential for any sound *mechanistic* interpretation.

7.2.4   Polyketides

Aided by modern spectroscopic, chromatographic and radioimmunological methods of detection and analysis, the use of cell-free systems provides a powerful tool for the recognition of biosynthetic intermediates, as the three examples above so clearly illustrate. The potential of this technique as a means of dissecting biosynthetic transformation to an even finer level is only just beginning to be explored. A significant number of biosynthetic processes, particularly the elaboration and cyclization of polyketides, are believed to take place under the control of multienzyme systems. The formation of as simple a metabolite as methylsalicylic acid (*7.21*) from three units of malonyl-CoA and an acetyl-CoA starter unit may require a minimum of six discrete steps, each leading to the transient generation of a distinct ground-state organic molecule. Direct observation of any of these individual transformations is complicated, however, because none of these intermediates ever leaves the active site of the multienzyme system, in this case presumably because the growing polyketide chain remains covalently tethered to the enzyme system by a thioester linkage analogous to that found in fatty acid synthetases. Methylsalicylate synthetase is in fact one of the few polyketide synthetases to have been even partially characterized (Scott *et al.*,

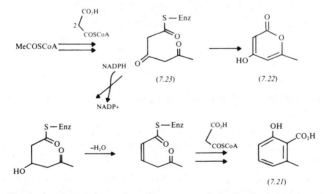

Fig. 7.9 Enzymic conversion of acetate to methylsalicylic acid (*7.21*), and formation of triacetic acid lactone (*7.22*) as a derailment metabolite.

CH₃COSCoA,   CH₃CH₂COSCoA,   CH₃CH₂CH₂COSCoA

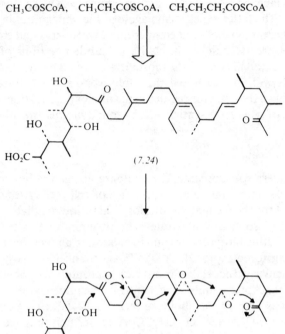

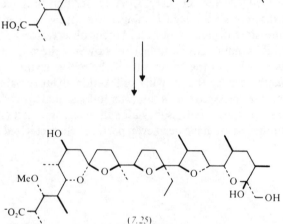

Fig. 7.10 Hypothetical derivation of monensin A (*7.25*) from acyclic triene (*7.24*).

1974; Dimroth *et al.*, 1976). Our understanding of this system relies heavily on analogy to the far more thoroughly studied prokaryotic and eukaryotic fatty acid synthetases (Volpe and Vagelos, 1973; Bloch and Vance, 1977). One of the few pieces of experimental information about the chain-building sequence comes from the observation that omission of NADPH from the incubation medium leads to the derailment product triacetic acid lactone

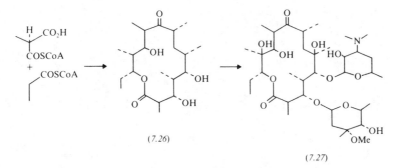

Fig. 7.11 Conversion of propionate to erythromycin A (7.27) via 6-deoxyery-thronolide B (7.26).

(7.22), which suggests the reasonable but still tentative conclusion that reduction of the diketoester (7.23) precedes condensation with the last malonate unit (Fig. 7.9) (Dimroth et al., 1976). The formation of more complex polyketides such as macrolides and polyethers present an experimental problem which is an order of magnitude more complicated. The conversion of the simple substrates acetate, propionate and butyrate to the triene (7.24), a suggested precursor of the anti-coccidial polyether, monensin (7.25) (Cane et al., 1981, 1982a), may require a minimum of 43 distinct condensation, reduction, dehydration and reduction steps (Fig. 7.10). There are currently no satisfactory experimental models to describe the packaging of an enzyme system responsible for as many distinct steps under such tight regio- and stereo-chemical control, or even to suggest how a substrate might be physically processed by such a large system. A similar problem is encountered in the conversion of propionate and methylmalonate to 6-deoxyerythronolide B (7.26), the earliest established polyketide precursor of the macrolide antibiotic erythromycin A (7.27) (Fig. 7.11) (Corcoran, 1981). Although the enzymes responsible for several of the late-stage oxidations and methylations involved in the conversion of (7.26) to (7.27) have been isolated and characterized, repeated attempts to observe cell-free formation of 6-deoxyerythronolide B have all met with failure, and in fact no intermediates between methylmalonate and (7.26) have ever been observed in either intact Streptomyces cultures or derived broken-cell preparations. The pessimistic conclusion of Corcoran that erythromycin synthetase may be merely a fatty acid synthetase responding to substrate pressure is almost certainly unwarranted. Isolation of the appropriate multi-enzyme systems remains a considerable challenge and may require new and more subtle methods of cell-disruption and manipulation of the resulting homogenates. One particularly promising approach to the study of multi-enzyme synthetases may be the application of modern recombinant DNA

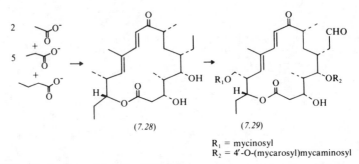

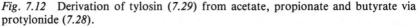

R_1 = mycinosyl
R_2 = 4'-O-(mycarosyl)mycaminosyl

Fig. 7.12 Derivation of tylosin (*7.29*) from acetate, propionate and butyrate via protylonide (*7.28*).

techniques (Sinsheimer, 1977; Wu, 1979) for isolation and eventual amplification of the appropriate genes. Work is already underway in several laboratories, including those of Baltz (Baltz, 1978; Baltz and Seno, 1981, Baltz *et al.*, 1981), Hutchinson (Wang *et al.*, 1982) and Omura (Omura *et al.*, 1981; Ikeda *et al.*, 1982), on the genetics of macrolide biosynthesis. (See also Hopwood and Merrick, 1977.) Baltz has already reported the interesting observation that from among 75 mutants of *S. venezuelae*, which had lost the ability to produce the 16-membered macrolide tylosin (*7.29*), 57 appeared to be blocked in the formation of the parent lactone system, protylonide (*7.28*) (Fig. 7.12). Baltz has suggested that the high incidence of this particular phenotype indicates that the affected genes code for a multienzyme synthetase which is unable to assemble if one of the component proteins is defective. Consistent with the existence of a multienzyme synthetase is the observation that none of the protylonide-negative mutants can act as secretors in co-synthesis assays, suggesting the absence of any diffusible intermediates. Further progress in this area will require the development of suitable plasmid or phage-derived cloning vectors, a goal which will no doubt be achieved within the next few years (Chater *et al.*, 1982).

7.2.5   Problems and prospects*

While work with cell-free systems provides significant new opportunities for biosynthetic inquiry, it must also be recognized that routine application of this tool is impeded by a number of severe problems. Over the years, secondary metabolic enzymes have acquired a reputation of being unusually difficult to isolate and handle. This folklore may account, in part, for the frequent reluctance of chemists to attempt to develop cell-free alternatives to traditional whole-cell feeding techniques. Moreover, the enormous difficulties encountered by experienced enzymologists in obtaining viable

* Further discussion of the relevant biochemistry can be found in Chapter 8.

penicillin-synthesizing preparations do little to dispel the notion that such enzymes are inherently unstable or require special techniques for their isolation. In most cases, however, the unsavoury reputations of secondary metabolic enzymes are probably undeserved. The use of polypyrollidone additives to neutralize potentially harmful phenolics, prior washing of cells with saline to reduce the titre of deleterious proteases, and the use of poly-styrene resins to remove endogenous lipophilic metabolites have in many instances alleviated the more common problems of rapid enzyme inactivation or degradation. As a group, the enzymes of secondary metabolism are probably no more intractable or subject to immediate destruction in cell homogenates than any other enzymes. The small number of terpenoid-cyclizing enzymes which have been examined to date are soluble and most are of moderate molecular weight (40 000—100 000) (Croteau, 1981; West, 1981) and are certainly easier to handle than many experimentally demanding but far better studied membrane-bound or glycoprotein systems. In fact the major problem presented by secondary metabolic enzymes is altogether different but nonetheless severe: they do not occur in very high intracellular concentrations and the reactions they catalyse are not particularly fast. For example, bornyl pyrophosphate synthetase, an enzyme which is discussed in more detail below, has a $V_{max.}$ for geranyl pyrophosphate of 11.4 nmol per mg of protein per hour (Croteau and Karp, 1979). On the basis of a molecular weight of 95 000 and assuming a minimum purity of 10% for this enzyme, a turnover number of 10 mol of substrate per mol of enzyme per hour can be calculated. Such extremely modest turnover rates are probably typical of terpenoid-cyclizing enzymes, although few data are as yet available on which to base a secure judgement. It is an accepted tenet of enzyme purification that the greater the concentration of a desired enzyme at the start of a purification sequence, the easier will be the eventual purification. In the extreme, the success of recombinant DNA methods in providing highly purified enzymes is based on magnification of physiological titres by two or three orders of magnitude, rendering subsequent purification an almost trivial operation (see, for example, Schleif and Favreau, 1982). More traditional methods of strain selection ordinarily achieve more modest results, but are based on the same principle of selective enzyme enrichment. The magnitude of the obstacle facing investigators wishing to purify secondary metabolic enzymes may be appreciated if one considers for a moment penicillin synthetase. Commercial strains of penicillin-producing organisms represent the extreme in strain selection for secondary metabolic activity, resulting in yields of 10—20 g or more per litre of culture. Such yields appear more modest, however, when one calculates that 10 g in *ca.* 5 days corresponds to a mere 80 mg per hour, assuming minimal turnover of the end product. Stated differently, the total penicillin synthetase in 1 litre of cultured cells is turning over something less than

1 μmol of product per second. Such calculations are very crude but are probably good to within an order of magnitude. What they reveal, however, is that typical isolation procedures for crude penicillin synthetase (O'Sullivan *et al.*, 1979; Meesschaert, 1980) are already yielding a reasonable proportion, perhaps as much as 10%, of the total enzyme available within the cell. The fact is that a litre of microbial culture accumulating 2 g of metabolite in a day, as impressive as that may be, cannot be compared with a litre of digestive fluid containing extracellular proteases which are hydrolysing kilograms of substrate! The problem becomes even more severe when one is forced to deal with cultures producing secondary metabolites at far more modest levels, 10–100 mg per litre. A major part of the problem has to do with the low turnover numbers. Most methods of enzyme separation are based on the characteristic properties of proteins; on the other hand most methods of enzyme detection are based on the effect of that protein on some substrate. The less enzyme one has or the slower a given quantity of enzyme processes its characteristic substrates, the lower are the limits necessary to detect and assay that enzyme in the first place. These limitations in enzyme activity are a common obstacle which must be overcome for any successful programme of enzyme purification.

Another problem that must be dealt with in choosing to work with cell-free systems is that the details of many of the most important biosynthetic *pathways* are still largely undefined. It should be obvious that in order to study any enzyme system one should know at least either the substrate or the product of the enzyme-catalysed reactions themselves, and preferably both. One of the most important contributions of stable-isotope NMR methods is that primary precursor–product relationships can be established relatively rapidly and, at the very least, educated guesses made about the general outlines of the pathway linking precursor to product. In some cases crude cell-free preparations may be found which support the entire sequence of transformations linking the primary precursors to the fully functionalized natural product. In such cases it may be possible to study shorter and shorter segments of a pathway until individual metabolic transformations come into focus. Frequently, however, one or more enzymes of the complete pathway are absent from a crude cell-free extract, for example those enzymes catalysing activation of precursors or mediating late-stage oxidations, methylations or glycosylations. Thus it is possible to have a viable cell-free preparation in hand which catalyses, for example, the cyclization of farnesyl pyrophosphate to some sesquiterpene hydrocarbon, but is unable to convert that hydrocarbon to its better known, more oxidized, metabolites. In the latter instance it is useful to have an authentic sample, either natural or synthetic, of the sesquiterpene in question, or some suitable analytical technique must be devised for detecting and identifying minute quantities of the material being generated.

There is no doubt that as experience is gained in working with the enzymes of secondary metabolism, many of the purely technical obstacles will be overcome by improved techniques of cell disruption, protein separation, including affinity methods, and substrate assay. As already mentioned, recombinant DNA methods for gene cloning and amplification hold the promise of significantly increasing the quantities of key enzymes which can be made available for study. Major problems in devising suitable methods for transformation and gene expression in eukaryotic systems from fungi to higher plants must be overcome before this methodology becomes generally useful. Nonetheless, these problems are currently under intensive study by molecular geneticists in both academic and industrial research laboratories (Hofschneider and Goebel, 1982), and the study of natural products biosynthesis is certain to benefit from advances made in this revolutionary field.

7.3 CELL-FREE STUDIES OF ISOPRENOID BIOSYNTHESIS

The preceding discussion will give some idea of the state of enzyme-level biosynthetic studies as of the middle of 1982. It will be recognized, of course, that for many years well-studied enzymes such as alcohol dehydrogenase, fumarase and malate synthetase, among others, have been employed as powerful tools for the synthesis or analysis of stereospecifically labelled substrates or degradation products (Arigoni and Eliel, 1969; see also Chapter 4). The present discussion has purposely been limited to a consideration of studies explicitly involving enzymes responsible for the formation of natural products themselves. In the following section is presented a more detailed description of current work in our own laboratories on enzymes of the terpenoid-biosynthetic pathway.

For the last few years we have been investigating the biosynthesis of a number of terpenoid metabolites produced by a variety of bacterial, fungal and higher-plant systems. In most cases the early stages of this work have been carried out by feedings to intact organisms and analysis of the resulting labelled metabolites by traditional chemical degradative techniques or by application of modern spectroscopic methods including high-field [^{13}C]- and [^{2}H]-NMR. Throughout much of this work we have been concerned with the metabolism of allylic pyrophosphates (Cane, 1980). The latter group of compounds plays a central role in the biosynthesis of isoprenoid metabolites and a study of their transformations is critical to a detailed understanding of the mechanisms of formation of this family of compounds. A good deal of our more recent work in this area has been carried out at the cell-free level, in some cases with enzymes already described in the recent literature, and in others with crude preparations which we have developed in our own laboratories. At this stage of our investigations we have been less concerned with

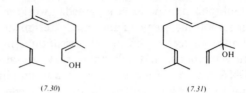

Fig. 7.13 Farnesol (*7.30*) and nerolidol (*7.31*).

the properties of these terpenoid-metabolizing enzymes as either proteins or catalytic entities than with defining as closely as possible the characteristic changes in structure and bonding undergone by each individual substrate in the course of its enzyme-mediated transformation.

The role of the commonly occurring allylic alcohols farnesol (*7.30*) and nerolidol (*7.31*) as precursors of sesquiterpene metabolites was first suggested some sixty years ago by Ruzicka (Fig. 7.13) (Ruzicka and Stoll, 1922). This proposal was reformulated in mechanistic terms in the early 1950s as the Biogenetic Isoprene Rule (Ruzicka, 1959, 1963) and can be invoked to account for the formation of now more than 200 individual sesquiterpene carbon skeletons (Hendrickson, 1959; Parker *et al.*, 1967). The identification of the activating group as a pyrophosphate ester was made in 1958 (Lynen *et al.*, 1958, 1959; Chayken *et al.*, 1958) in the course of studies on sterol biosynthesis. Although farnesyl pyrophosphate itself, or in some cases, free farnesol, has been incorporated into a small number of sesquiterpene metabolites by feedings to intact plants or microbial cultures, most of the evidence for the role of this key intermediate has been deduced from the results of feeding earlier precursors, such as labelled acetate or mevalonate, and analysis of the isotopic distribution in the derived labelled metabolites (Cane, 1981). In spite of the wealth of valuable information which has been provided by these experiments, a number of essential mechanistic features of the key cyclization reactions have been difficult to establish. One of the most vexing problems has come from the realization that the formation of six-membered rings from a *trans*-allylic pyrophosphate requires isomerization

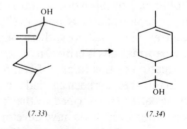

Fig. 7.14 Cyclization of linalool to α-terpineol.

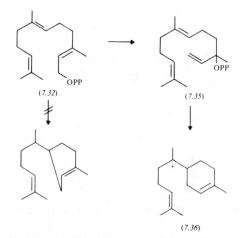

Fig. 7.15   Formation of bisabolyl cation (*7.36*) from farnesyl pyrophosphate (*7.32*) by initial isomerization to nerolidyl pyrophosphate (*7.35*).

of the *trans*-2,3-double bond of the precursor in order to avoid the formation of a prohibitively strained *trans*-cyclohexene. The recognition of this conceptual difficulty led to the generation of two conflicting theories to account for the isomerization–cyclization of the initially formed substrate *trans,trans*-farnesyl pyrophosphate (*7.32*). The first of these hypotheses, based on the long-recognized ability of activated derivatives of linalool (*7.33*), a tertiary allylic alcohol, to cyclize to α-terpineol (*7.34*) (Fig. 7.14) (Stephan, 1898), envisioned initial allylic isomerization of (*7.32*) to nerolidyl pyrophosphate (*7.35*) which, following rapid rotation about the newly generated 2,3-single bond, can undergo facile cyclization (Fig. 7.15). The alternative scheme for double-bond isomerization was inspired by the observation that farnesal (*7.37*), the aldehyde derived from farnesol, undergoes rapid *cis,trans* isomerization at physiological pH and temperature. It was therefore suggested that the cyclization of *trans,trans*-farnesyl pyrophosphate might be explained by initial hydrolysis of the pyrophosphate ester to the free alcohol followed by oxidation to farnesal with concomitant loss of one of the C-1 hydrogen atoms. Isomerization of the farnesal with subsequent reduction and repyrophosphorylation would give *cis,trans*-farnesyl pyrophosphate which is geometrically competent to cyclize (Fig. 7.16). Analogous schemes had also been advanced to account for the formation of cyclic monoterpenes from the accepted universal precursor geranyl pyrophosphate. The evidence bearing on both these theories was reviewed in 1980 (Cane, 1980) and need not be considered here. It is sufficient to note that within the last three or four years experiments carried out by several independent laboratories, for the most part utilizing cell-free systems, have conclusively excluded all redox mechanisms proposed to date and are completely

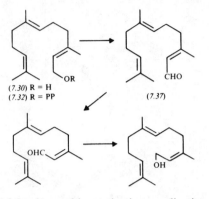

Fig. 7.16 Redox model for farnesol isomerization–cyclization.

consistent with the alternative route involving the intermediacy of the tertiary allylic isomer, nerolidyl pyrophosphate. Some of our own contributions to the solution of this problem are described below.

7.3.1 Enzymic isomerization of farnesyl to nerolidyl pyrophosphate

Although the probable connection between farnesyl and nerolidyl pyrophosphate had been recognized for some time, until 1978 opportunities for studying this important transformation had been limited by the absence of appropriate biochemical systems known to catalyse the allylic isomerization reaction. A suitable enzyme preparation first became available when we found that a cell-free system from the fungus *Gibberella fujikuroi* was capable of converting both farnesyl pyrophosphate and nerolidyl pyrophosphate to the cyclopentanoid metabolite cyclonerodiol (Cane and Iyengar,

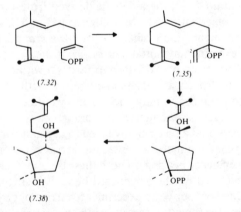

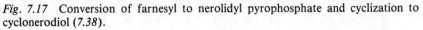

Fig. 7.17 Conversion of farnesyl to nerolidyl pyrophosphate and cyclization to cyclonerodiol *(7.38)*.

1978; Cane *et al.*, 1981b). The actual product of the addition of water across the vinyl and central double bonds of nerolidyl pyrophosphate turned out to be the corresponding pyrophosphate ester of cyclonerodiol which is hydrolysed to cyclonerodiol itself by a separable phosphatase with cleavage of the P–O bond (Fig. 7.17). The specificity of the cyclization was confirmed by using [12,13-^{14}C]nerolidyl pyrophosphate as substrate and degrading the resulting cyclonerodiol to establish the presence of all the label in the allylic methyl groups of the side chain. Using partially purified enzymes it was possible to observe the cyclization of [1,2-^{13}C$_2$]nerolidyl pyrophosphate directly by [^{13}C]-NMR by monitoring the appearance of the coupled doublets ($J = 38.4$ Hz) corresponding to C-1 and C-2 of the product. Moreover, the cyclase was shown to be specific for the (3R)-enantiomer of nerolidyl pyrophosphate, while incubation of (E)-[1,2-^{2}H$_2$,1-^{3}H]nerolidyl pyrophosphate and determination of the chirality of the resulting labelled C-1 methyl group of cyclonerodiol established that the addition of water across the vinyl and central double bonds of nerolidyl involves *trans,trans* stereochemistry, as illustrated in Fig. 7.18. Competitive incubation experiments with a phosphatase-free enzyme preparation which had been purified some tenfold by a sequence of ammonium sulphate precipitation, Sephadex G-150 chromatography and treatment with hydroxylapatite, established that nerolidyl pyrophosphate was converted to cyclonerodiol approximately five times more rapidly than was its precursor, farnesyl pyrophosphate.

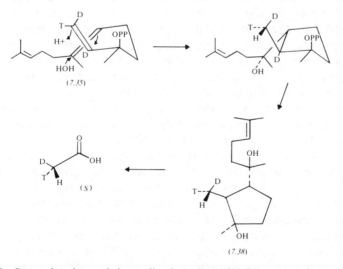

Fig. 7.18 Stereochemistry of the cyclization of nerolidyl pyrophosphate to cyclonerodiol.

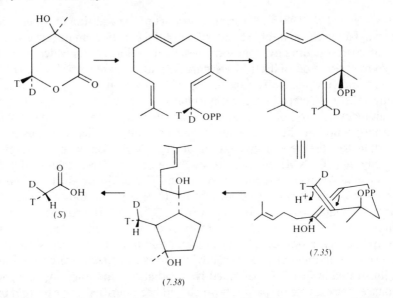

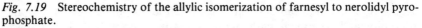

Fig. 7.19   Stereochemistry of the allylic isomerization of farnesyl to nerolidyl pyrophosphate.

    The cell-free system was also used to study the farnesyl to nerolidyl pyro-
phosphate conversion itself (Fig. 7.17). Short-term (15 min) incubation of
[12,13-^{14}C]farnesyl pyrophosphate gave labelled nerolidyl pyrophosphate
which was isolated after addition of inactive carrier and degraded to show
that all of the ^{14}C activity was at the expected sites, C-12 and C-13. With the
knowledge of the stereochemical course of the protonation at C-1 of
nerolidyl pyrophosphate already gained from the foregoing cell-free studies,
feeding stereospecifically deuterated and tritiated mevalonates to whole cells
of *G. fujikuroi* and subsequent chiral methyl group analysis of the derived
cyclonerodiol samples unambiguously established that the enzymic rearrangement of farnesyl pyrophosphate to its tertiary allylic isomer, nerolidyl
pyrophosphate, takes place, not surprisingly, with exclusive *syn* stereochemistry (Fig. 7.19) (Cane *et al.*, 1978, 1981b).
    Having found that the farnesyl–nerolidyl pyrophosphate rearrangement
takes place on a single face of the allylic system, we turned our attention to
the role of the pyrophosphate moiety during the isomerization, in the hopes
of gaining some insight into the timing of the bond-breaking and bond-
making steps (Cane and Iyengar, 1979; Cane *et al.*, 1981b). In designing our
experiments, we considered four idealized mechanisms which might account
for the observed allylic rearrangement (Fig. 7.20): A, a concerted phospho-
Claisen or ionic stepover mechanism in which one of the two enantiotopic
non-bridge oxygens attached to the proximal phosphorus atom becomes

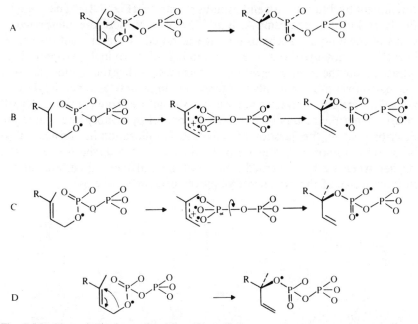

Fig. 7.20 Hypothetical mechanisms for allylic pyrophosphate rearrangement. A, Phospho-Claisen rearrangement; B, free pyrophosphate ion; C, tight ion-pair; D, 1,3-rearrangement.

attached to C-3 in the tertiary allylic pyrophosphate product; B, a stabilized allylic cation, or its covalently bound equivalent, in which free inorganic pyrophosphate is formed, allowing complete scrambling of all six non-bridge oxygens; C, an ion-pair intermediate in which there is sufficient time for equilibration of the three temporally equivalent oxygen atoms attached to the proximal phosphorus; and D, a formal 1,3-sigmatropic rearrangement or tight ion-pair in which P_{α}-OP rotation is restricted, resulting in exclusive recapture by the allylic substrate of the original ester oxygen atom. These mechanisms were readily distinguished by incubation of [1-^{18}O]farnesyl pyrophosphate with partially purified enzyme preparations from *G. fujikuroi* and determination of the [18]O content of the product. In using this enzyme system we took advantage of the *in situ* conversion of the initially formed nerolidyl pyrophosphate to the unreactive end product cyclonerodiol, whose C-3 oxygen atom we had previously shown was identical to the corresponding C-3 ester oxygen of nerolidyl pyrophosphate.

Mass-spectroscopic analysis of a derivative of cyclonerodiol, the tri-methylsilyllactone (*7.39*), established that only one-third of the original [18]O of the precursor was present at C-3 of the product while control experiments ruled out either prior scrambling of the precursor or reduction of the [18]O enrichment by dilution with endogenous substrates (Fig. 7.21). These results firmly excluded mechanisms A, B and D and were therefore interpreted in terms of the ion-pair mechanism C. It was noted, however, that the experiment had measured only net scrambling of nerolidyl pyrophosphate released from the isomerase without determining either the scrambling for a single turnover nor the number of times bound farnesyl and nerolidyl pyrophosphate are interconverted at the active site before product release. As will be seen from the results of the study of bornyl pyrophosphate synthetase described below, the question of active-site equilibration has turned out to be of some importance and we plan to examine the allylic isomerization further when we have learned how to obtain sufficient quantities of the isomerase free of the cyclonerodiol pyrophosphate synthetase.

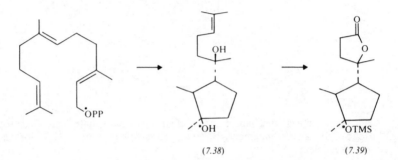

(7.38)   (7.39)

Fig. 7.21   Conversion of $[1 - {}^{18}O]$farnesyl pyrophosphate to $[3 - {}^{18}O]$cyclonerodiol.

7.3.2   *Isomerization–cyclization of farnesyl pyrophosphate. Trichodiene biosynthesis*

The experiments on the enzymic conversion of farnesyl to nerolidyl pyrophosphate provided information on both the stereochemical course of the rearrangement and the timing of critical bond-breaking and bond-making events leading to a proposed ion-pair mechanism. In fact, this ion-pair mechanism can be invoked to explain not only simple allylic isomerizations, but more complex isomerization–cyclizations to form six-membered rings as well (Cane *et al.*, 1981b). Towards the end of our work on farnesyl pyrophosphate isomerase we had begun to examine a second fungal enzyme system, first described by Hanson, which catalyses the cyclization of farnesyl pyrophosphate to trichodiene (*7.40*) (Evans and Hanson, 1976), the parent hydrocarbon (Nozoe and Machida, 1972) of the trichothecane family of sesquiterpene antibiotics (Tamm and Breitenstein, 1980; Cane, 1981). It had also been reported that incubation of $[1,5,9-{}^{3}H_6,4,8,12-{}^{14}C_3]$*trans,trans*-farnesyl pyrophosphate (${}^{3}H/{}^{14}C$ atom ratio 6:3) with the cell-free system

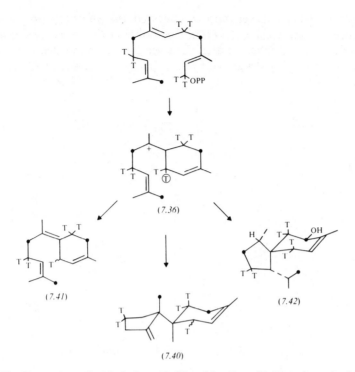

Fig. 7.22 Formation of γ-bisabolene (*7.41*), trichodiene (*7.40*) and coccinol (*7.42*) from the common bisabolyl (*7.36*) intermediate.

from *Trichothecium roseum* gave trichodiene with concurrent loss of a C-1 hydrogen atom of farnesyl pyrophosphate, based on the measured $^3H/^{14}C$ atom ratio of 5:3 (Fig. 7.22). These findings appeared to support a redox mechanism for the isomerization−cyclization sequence. Unfortunately this conclusion had not been substantiated by appropriate degradations to determine the distribution of isotopic label nor had suitable derivatives of substrates and products been recrystallized to constant activity. Interestingly, similar findings had been made earlier by Overton who reported that conversion of $[5-^3H_2,2-^{14}C]$mevalonate to (Z)-γ-bisabolene (*7.41*) by a cell-free system obtained from tissue cultures of *Andrographis paniculata* resulted in loss of one-sixth of the tritium label, presumably corresponding to one of the C-1 hydrogen atoms of the intermediate $[1,5,9-^3H_6,4,8,12-^{13}C_6]trans,trans$-farnesyl pyrophosphate (Overton and Picken, 1976; Mackie and Overton, 1977). Curiously, however, the trichodiene synthetase appeared to be specific for removal of 1-H_{si} of farnesyl pyrophosphate whereas loss of 1-H_{re} was implicated in the action of bisabolene synthetase, in spite of the fact that both cyclizations presumably involve an intermediate bisabolyl cation (*7.36*).

To make the situation even more complicated, the whole cell biosynthesis of the sesquiterpene coccinol (*7.42*), a metabolite of *Fusidium coccineum* also derivable from a bisabolyl cation, had been shown quite clearly to involve loss of neither C-1 hydrogen of farnesyl pyrophosphate (Gotfredsen, 1978; Cane, 1981).

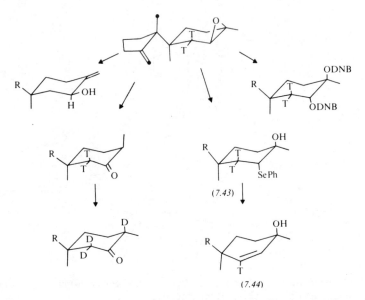

Fig. 7.23 Degradation of trichodiene derived from $[1 - {}^3H_2, 12, 13 - {}^{14}C]$farnesyl pyrophosphate.

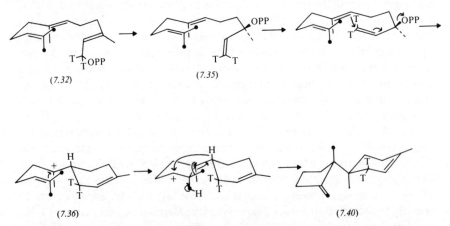

Fig. 7.24 Mechanism of isomerization–cyclization of farnesyl pyrophosphate to trichodiene.

We therefore chose to re-examine the trichodiene synthetase reaction in the hope of resolving some of these apparent contradictions (Cane *et al.*, 1981c). Initial small-scale incubations using [1-^{3}H,12,13-^{14}C]farnesyl pyrophosphate as substrate revealed that neither the efficiency of conversion to trichodiene nor the ^{3}H/^{14}C ratio of the cyclization product were affected by the presence or absence of a mixture of NAD$^+$, NADH and NADPH, an observation inconsistent with a redox isomerization mechanism. Most importantly, preparative-scale incubation of the [1-^{3}H,12,13-^{14}C]farnesyl pyrophosphate (^{3}H/^{14}C atom ratio 2:2) and degradation of the resulting trichodiene by the sequence of reactions illustrated in Fig. 7.23 established unambiguously that cyclization of *trans,trans*-farnesyl pyrophosphate to trichodiene took place without loss of hydrogen from C-1 of the precursor. These results firmly excluded any redox mechanism and strongly favoured the alternative isomerization—cyclization pathway involving the tertiary allylic isomer nerolidyl pyrophosphate (Fig. 7.24). In order to test this scheme further, we have determined the stereochemical course of the cycliz-ation by utilizing (1S)-[1-^{3}H,12,13-^{14}C]farnesyl pyrophosphate and subject-ing the derived trichodiene to an analogous degradative sequence. The absence of any tritium from the allylic alcohol (*7.44*) obtained by *syn* elimi-nation of the selenoxide derived from (*7.43*), established that the tritium label in trichodiene resided exclusively in the *α*-position, implying that the cyclization of *trans,trans*-farnesyl pyrophosphate took place with net *retention* of configuration at C-1 (Fig. 7.25). This result was completely consistent with the mechanism of Fig. 7.24 as well as the previously estab-lished *syn* stereochemistry of the farnesyl—nerolidyl isomerization. It is believed that the entire sequence takes place at the active site of a single enzyme and the stage is now set to test the intermediacy of nerolidyl pyro-phosphate directly, as well as to determine its conformation in the key cycliz-ation reaction. The availability of a cell-free system has also provided the opportunity to try to uncouple the otherwise cryptic isomerization reaction from the tightly coupled cyclization step by the use of appropriately designed substrate analogues. In fact, the complexity of the transformations believed to be catalysed by trichodiene synthetase, involving several distinct bond-making and bond-breaking steps and at least one ground-state intermediate,

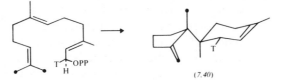

(*7.40*)

Fig. 7.25 Conversion of (1S)-[1 – ^{3}H,12,13 – ^{14}C]farnesyl pyrophosphate to trichodiene with retention of configuration at C – 1.

is probably typical of allylic pyrophosphate cyclases in general. A major challenge in future work with enzymes of this class will be to devise ways to dissect such multistep transformations so as to allow direct observation of transiently generated intermediates which never escape the active site under normal conditions, as well as to find methods to probe the manifold of cationic structures generated during the enzyme-mediated rearrangement. One of the most intriguing problems which must be confronted is to make a distinction between the role of these enzymes in catalysing the reaction of the common substrate, farnesyl pyrophosphate (rate enhancement), and in controlling the exclusive formation of a single one of a multitude of possible products (specificity). Although it is much too early to reach any reliable conclusions, it is interesting to note that while the reactions catalysed by these enzymes do not appear to be particularly fast (turnover numbers may range between 0.001 and $1.0 \, s^{-1}$) (cf. Laskovic and Poulter, 1981), the Michaelis constants for substrate are very respectable ($1-50 \, \mu M$). In the meantime the overall mechanistic picture has already become a good deal more coherent, as Overton has re-examined his initial experiments with bisabolene synthetase using $[1-^3H_2,12,13-^{14}C]$farnesyl pyrophosphate as substrate and has now concluded that, in this system too, cyclization proceeds without loss of isotope from C-1 of the farnesyl precursor (Anastastis *et al.*, 1982).

7.3.3   *Monoterpene biosynthersis. Bornyl pyrophosphate synthetase*

During the same period in which the above investigations were being carried out, remarkable progress was being made in the closely related area of monoterpene biosynthesis by Croteau's group at Washington State University (Croteau, 1981). The majority of the known monoterpenes are produced by higher plants, and over the years associated problems of low incorporations, compartmentalization effects, and rapid deactivation of cell-free preparations, exacerbated by an all-too-frequent lack of experimental rigour had combined to generate a virtual quicksand of dubious claims and questionable hypotheses to explain the formation of this deceptively simple family of compounds. Over the last few years, working with cell-free preparation from a small group of plants known for their production of essential oils, including sage, mint, tansy, and fennel, Croteau has made major advances in understanding the formation of the main classes of monocyclic and bicyclic monoterpenes, including *p*-menthanes, pinanes, bornanes, fenchanes and thujanes. Much of this extensive work has been reviewed elsewhere (Croteau, 1981) and will not be considered here. It is important to note, however, that cyclization of geranyl pyrophosphate to representatives of each of the major monoterpene families has been shown beyond any doubt to take place without loss of hydrogen from C-1 (Croteau and Felton,

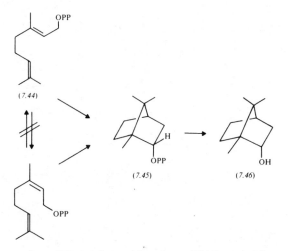

Fig. 7.26 Conversion of geranyl pyrophosphate (*7.44*) to (+)-bornyl pyrophosphate (*7.45*).

1981) and that none of these enzyme-catalysed cyclizations require nicotina-mide coenzymes (see, for example, Gambliel and Croteau, 1982). One particularly exciting observation was the finding that an enzyme system from sage (*Salvia officinalis*) catalyses the conversion of geranyl pyrophosphate (*7.44*) to (+)-bornyl pyrophosphate (*7.45*), which is subsequently hydro-lysed by a distinct pyrophosphatase to (+)-borneol (*7.46*) (Fig. 7.26) (Croteau and Karp, 1977). Croteau also found that in the absence of con-taminating phosphatases and pyrophosphatases which compete for the allylic pyrophosphate substrates, geranyl pyrophosphate was the preferred substrate for cyclization, with a V_{max}/K_m 20 times that of its *cis* isomer, neryl pyrophosphate (Croteau and Karp, 1979). Not only could neither of these observations have been possible using intact plants, but the finding that the cyclization product is itself a pyrophosphate ester provided a unique opportunity to examine experimentally the role of the pyrophosphate moiety in the coupled isomerization–cyclization process. In collaboration with Professor Croteau we therefore undertook [^{18}O]-labelling studies analogous to those performed earlier in our study of farnesyl pyrophosphate isomeriz-ation (Cane et al., 1982b).

A reasonable model for the formation of (+)-bornyl pyrophosphate from geranyl pyrophosphate, analogous to the proposed model of trichodiene formation and based on closely related chemical models (Gotfredsen et al., 1977; Gotfredsen, 1978) is illustrated in Fig. 7.27. Initial isomerization of geranyl pyrophosphate to its tertiary allylic isomer followed by 2,3-single bond rotation would generate linalyl pyrophosphate (*7.47*) in a *cisoid* conformation suitable for ionization and electrophilic attack on the

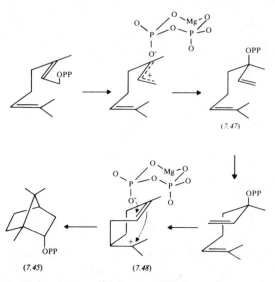

Fig. 7.27 Mechanism of isomerization–cyclization of geranyl pyrophosphate (7.44) to (+)-bornyl pyrophosphate.

6,7-double bond. The transiently generated α-terpinyl cation (7.48), formally analogous to the bisabolyl intermediate (7.36) of sesquiterpene cyclizations, can further cyclize by attack on the newly formed cyclohexene double bond and capture of the resultant cation by the paired inorganic pyrophosphate ion. Of particular interest was the extent to which the pyrophosphate moiety becomes free of its cationic partner during the course of the cyclization.

The first indication that the two ends of the pyrophosphate retained their identity during the formation of bornyl pyrophosphate came from experiments with [^{32}P]-labelled substrates (Fig. 7.28). Separate incubations of [1-^{3}H$_2$,α-^{32}P]- and [1-^{3}H,β-^{32}P]-geranyl pyrophosphate with phosphatase-free preparations of (+)-and (−)-bornyl pyrophosphate synthetases from *Salvia officinalis* (sage) and *Tanacetum vulgare* (tansy) gave [^{3}H,^{32}P]bornyl pyrophosphates which were selectively hydrolysed to the corresponding bornyl phosphates. Measurement of the ^{3}H/^{32}P ratio established that only samples of (+)- and (−)-bornyl phosphate derived from [1-^{3}H$_2$,α-^{32}P] geranyl pyrophosphate retained ^{32}P thereby ruling out tumbling of the pyrophosphate moiety during the cyclization. Use of ^{18}O-labelled substrates then revealed an unexpectedly rigid binding of the inorganic pyrophosphate anion–terpenoid cation pair at the active site of the cyclase. Thus incubation of [1-^{18}O]geranyl pyrophosphate with (+)-bornyl pyrophosphate synthetase gave a product, analysed by mass spectrometry of the derived benzoate, which retained between 75 and 100% of the original ^{18}O isotope attached to C-2. Analogous incubations of [1-^{18}O]geranyl pyrophosphate with the corresponding (−)-bornyl pyrophosphate synthetase preparation and analysis of the derived (−)-bornyl benzoate confirmed the complete lack of positional oxygen isotope exchange during the cyclization, the product (−)-borneol having an ^{18}O enrichment essentially unchanged from that of the geranyl precursor.

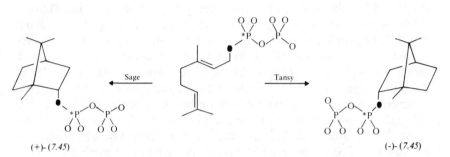

(+)-(7.45)   (-)-(7.45)

Fig. 7.28 Conversion of [1 − ^{18}O]geranyl [α − ^{32}P]pyrophosphates to (+)- and (−)-bornyl pyrophosphates by sage and tansy enzymes.

The remarkably tight restriction on the motion of the transiently generated inorganic pyrophosphate moiety is unexpected, particularly in the light of our earlier observation of the equilibration of the proximal phosphate oxygens during the closely related enzymic isomerization of farnesyl to nerolidyl pyrophosphate (Cane *et al.*, 1981b). While an analogous equilibrium between geranyl and linalyl pyrophosphate may exist at the active site of bornyl pyrophosphate synthethase, the initial cyclization step involving formation of a new C−C bond is most likely irreversible. The observed lack of positional isotope exchange is all the more striking when one considers the transient generation of an α-terpinyl cation−pyrophosphate anion pair in which the charge separation is at least 0.3 nm (3 Å). While it may be concluded that both the initial isomerization step as well as the subsequent cyclization are fast compared to positional isotope exchange, no information is available as to the rate of ring closure relative to either forward or reverse isomerization, nor is it yet possible to determine whether the restricted rotation is due to the interaction of the enzyme and the pyrophosphate moiety or to the inherently strong attraction of the inorganic pyrophosphate−Mg^{2+} complex for the paired carbocations. We do note, however, that Poulter has recently reported that [1-^{18}O]geranyl pyrophosphate reisolated from incubations with prenyl transferase has not undergone detectable scrambling, in spite of strong evidence for the generation of allylic cations at the enzyme active site (Mash *et al.*, 1981).

7.3.4   Cell free cyclization of farnesyl pyrophosphate to pentalenene

In addition to cyclizations proceeding through a bisabolyl cation or its equivalent, a large and varied group of sesquiterpenes are formed by initial cyclization of C-1 of farnesyl pyrophosphate to the distal double bond with generation of a 10- or 11-membered ring (Cane, 1981). We have been interested in the biosynthesis of several of these compounds containing a dimethylcyclopentane ring and formally derivable from humulene (7.49) (Cane and Nachbar, 1978; Cane et al., 1981d). Recently (Cane and Tillman, 1983) we have made an important advance in addressing the numerous biosynthetic problems presented by these metabolites by obtaining a cell-free extract of Streptomyces which catalyses the cyclization of trans,trans-farnesyl pyrophosphate to pentalenene (7.46) (Seto and Yonehara, 1980), the parent hydrocarbon of the pentalenolactone family of sesquiterpene antibiotics, whose mevalonoid origin we had previously established (Fig. 7.29) (Cane et al., 1981d). Incubation of [8-^{3}H,12,13-^{14}C]farnesyl pyrophosphate (^{3}H/^{14}C atom ratio 2:2) with the 34 000 g supernatant of a broken cell preparation of Streptomyces UC5319, obtained by rapid stirring with glass beads, gave labelled pentalenene. On the basis of the recovered ^{14}C activity, a turnover rate of 0.03 nmol of pentalenene per mg protein per hour was calculated. The recovered pentalenene was diluted with synthetic racemic carrier and a portion was converted to a mixture of diastereomeric cis-6,7-diols (7.47), each of which was separately recrystallized to constant activity. Both solid derivatives showed the expected loss of 50% of the original tritium (^{3}H/^{14}C atom ratio 0.9:2). Hydroboration–oxidation of the remaining pentalenene followed by pyridinium chlorochromate oxidation gave pentalen-7-one (7.48) which was essentially devoid of tritium, thereby locating all tritium in (7.46) at the expected site, C-7.

The observed labelling results are consistent with the pathway illustrated in Fig. 7.31 in which farnesyl pyrophosphate is first converted to humulene (7.49). Reprotonation at C-10 initiates a cyclization of humulene to the protoilludyl cation (7.50), which can undergo a hydride shift and further cyclization to pentalenene with concomitant loss of one of the protons originally at C-8 of farnesyl pyrophosphate. The cyclization stereochemistry illustrated has previously been established by whole-cell-feeding experiments

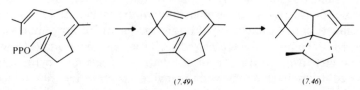

Fig. 7.29   Cyclization of farnesyl pyrophosphate to pentalenene (7.46) via humulene (7.49).

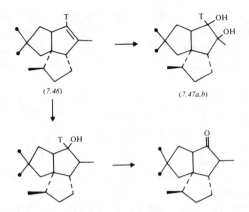

Fig. 7.30 Degradation of pentalenene derived from [8 – ³H,12,13 – ¹⁴C₂]farnesyl
pyrophosphate.

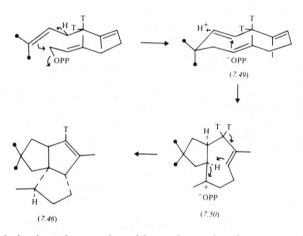

Fig. 7.31 Mechanism of conversion of farnesyl pyrophosphate to pentalenene.

(Cane *et al.*, 1981d), while the cyclization of the protoilludyl cation draws
analogy in the biomimetic synthesis of pentalenene from (7.50) reported
earlier by Shirhama and Matsumoto (Ohfune *et al.*, 1976). We have
previously suggested on stereochemical grounds that the enzymic formation
and further cyclization of the intermediate humulene take place at the same
active site (Cane *et al.*, 1981d). Circumstantial evidence consistent with this
hypothesis was the apparent absence of free humulene in the incubation
medium, as determined by attempted dilution trapping with unlabelled
carrier. With the availability of a defined enzyme system we hope to test these
notions further by determining whether the proton lost in the formation of

humulene can take part in the subsequent reprotonation which initiates further cyclization. The pentalenene synthetase reaction also provides a convenient vehicle to determine the stereochemistry of intramolecular S_E' reactions, of which there are two examples in the overall cyclization sequence.

The exploration of terpenoid enzymes has opened up vast new areas of inquiry, of which our own work is only a small part. With the use of cell-free systems, equally exciting opportunities are now apparent in the investigation of polyketide and alkaloid biosynthesis. These newly developed investigative tools will be most valuable if used not merely to solve old problems addressable by traditional whole-cell techniques, but to pose altogether new questions about the pathways used by Nature in synthesizing complex organic molecules and the marvellous catalysts which have been created to carry out the task.

REFERENCES

Aberhart, J. (1977) *Tetrahedron*, **33**, 1545.

Anastasis, P., Freer, I., Gilmore, C., Mackie, H., Overton, K. and Swanson, S. (1982) *J. Chem. Soc., Chem. Commun.*, 268.

Arigoni, D. and Eliel, E. L. (1969) *Top. Stereochem*, **4**, 127.

Arnstein, H. R. V. and Morris, D. (1960) *Biochem. J.*, **76**, 357.

Baldwin, J. E., Johnson, B. L., Usher, J. J., Abraham, E. P. Huddleston, J. A. and White, R. L. (1980a) *J. Chem. Soc., Chem. Commun.*, 1271.

Baldwin, J. E., Singh, P. D., Yoshida, M., Sawada, Y. and Demain, A. (1980b) *Biochem. J.*, **186**, 889.

Baldwin, J. E., Keeping, J. W., Singh, P. D. and Vallejo, C. A. (1981) *Biochem. J.*, **194**, 649.

Baltz, R. H. (1978) *J. Gen. Microbiol.*, **107**, 93.

Baltz, R. H. and Seno, E. T. (1981) *Antimicrob. Agents Chemother.*, **20**, 214.

Baltz, R. H., Seno, E. T., Stonesifer, J., Matsushima, P. and Wild, G. M. (1981) *Microbiology*, 371.

Battersby, A. R. (1967) *Pure Appl. Chem.*, **14**, 117.

Battersby, A. R. (1971) *The Alkaloids*, **1**, 31.

Battersby, A. R. and Gibson, K. H. (1971) *Chem. Commun.*, 902.

Battersby, A. R. and McDonald, E. (1975) in *Falk's Porphyrins and Metalloporphyrins*, 2nd edn (ed. K. M. Smith), Elsevier, Amsterdam, pp. 61–122.

Battersby, A. R., Brown, R. T., Kapil, R. S., Martin, J. A. and Plunkett, A. O. (1966) *Chem. Commun.*, 890.

Battersby, A. R., Kapil, R. S. and Southgate, R. (1968a) *Chem. Commun.*, 131.

Battersby, A. R., Kapil, R. S., Martin, J. A. and Mo, L. (1968b) *Chem. Commun.*, 133.

Battersby, A. R., Burnett, A. R. and Parsons, P. G. (1968c) *Chem. Commun.*, 1282.

Battersby, A. R., Burnett, A. R. and Parsons, P. G. (1969) *J. Chem. Soc. (C)* 1193.

Battersby, A. R., Ihara, M., McDonald, E., Satoh, F. and Williams, D. C. (1975) *J. Chem. Soc., Chem. Commun.*, 436.

Battersby, A. R., McDonald, E., Thompson, M. and Bykovsky, V. Ya. (1978) *J. Chem. Soc., Chem. Commun.*, 150.

Battersby, A. R., Matcham, G. W. J., McDonald, E., Neier, R., Thompson, M., Woggon, W.-D., Bykovsky, V. Ya. and Morris, H. R. (1979) *J. Chem. Soc., Chem. Commun.*, 185.

Battersby, A. R., Bushell, M. J., Jones, C., Lewis, N. G. and Pfenninger, A. (1981) *Proc. Natl. Acad. Sci. U.S.A.*, **78**, 13.

Battersby, A. R., Frobel, K., Hammerschmidt, F. and Jones, C. (1982) *J. Chem. Soc., Chem. Commun.*, 455.

Blackstock, W. P., Brown, R. T. and Lee, G. K. (1971) *Chem. Commun.*, 910.

Bloch, K. (1952) *Harvey Lect.*, **48**, 68.

Bloch, K. and Vance, D. (1977) *Annu. Rev. Biochem.*, **46**, 263.

Bogorad, L. (1966) in *The Chlorophylls* (eds L. P. Vernon and G. R. Seeley), Academic Press, New York, pp. 481–510.

Brechbueler-Bader, S., Coscia, C. J., Loew, P., von Szczepanski, Ch. and Arigoni, D. (1968) *Chem. Commun.*, 136.

Brown, S. A. (1972) *Biosynthesis*, 1, 1.

Burnham, B. F. (1969) in *Metabolic Pathways*, 3rd edn, Vol. III (ed. D. M. Greenberg), Academic Press, New York, pp. 403–537.

Cane, D. E. (1980) *Tetrahedron*, **36**, 1109.

Cane, D. E. (1981) in *Biosynthesis of Isoprenoid Compounds* (eds J. W. Porter and S. L. Spurgeon), John Wiley, New York, pp. 283–374.

Cane, D. E. and Iyengar, R. (1978) *J. Am. Chem. Soc.*, **100**, 3256.

Cane, D. E. and Iyengar, R. (1979) *J. Am. Chem. Soc.*, **101**, 3385.

Cane, D. E. and Nachbar, R. B. (1978) *J. Am. Chem. Soc.*, **100**, 3208. (corr. **101**, 1908).

Cane, D. E. and Tillman, A. M. (1983) *J. Am. Chem. Soc.*, **105**, 122.

Cane, D. E., Iyengar, R. and Shiao, M.-S. (1978) *J. Am. Chem. Soc.*, **100**, 7122.

Cane, D. E., Liang, T.-C. and Hasler, H. (1981a) *J. Am. Chem. Soc.*, **103**, 5962.

Cane, D. E., Iyengar, R. and Shiao, M.-S. (1981b) *J. Am. Chem. Soc.*, **103**, 914.

Cane, D. E., Swanson, S. and Murthy, P. P. N. (1981c) *J. Am. Chem. Soc.*, **103**, 2136.

Cane, D. E., Rossi, T., Tillman, A. M. and Pachlatko, J. P. (1981d) *J. Am. Chem. Soc.*, **103**, 1838.

Cane, D. E., Liang, T.-C. and Hasler, H. (1982a) *J. Am. Chem. Soc.*, **104**, 7274.

Cane, D. E., Saito, A., Croteau, R., Shaskus, J. and Felton, M. (1982b) *J. Am. Chem. Soc.*, **104**, 5831.

Chater, K. F., Hopwood, D. A., Kieser, T. and Thompson, C. J. (1982) in *Current Topics in Microbiology and Immunology. Gene Cloning in Organisms Other than E. coli*, (eds P. H. Hofschneider and W. Goebel), Springer-Verlag, Berlin, pp. 69–95.

Chayken, S., Law, J., Phillips, A. H., Tchen, T. T. and Bloch, K. (1958) *Proc. Natl. Acad. Sci., U.S.A.*, **44**, 998.

Clayton, R. B. (1965) *Q. Rev.*, **19**, 168.

Collie, J. N. (1893) *J. Chem. Soc.*, **63**, 329.

Corcoran, J. W. (1981) in *Antibiotics IV. Biosynthesis* (ed. J. W. Corcoran), Springer-Verlag, Berlin, pp. 132–174.

Cornforth, J. W. (1973) *Chem. Soc. Rev.*, **2**, 1.

Croteau, R. (1981) in *Biosynthesis of Isoprenoid Compounds* (eds J. W. Porter and S. L. Spurgeon), John Wiley, New York, pp. 225–282.

Croteau, R. and Felton, M. (1981) *Arch. Biochem. Biophys.*, **207**, 460.

Croteau, R. and Karp, F. (1977) *Arch. Biochem. Biophys.*, **184**, 77.

Croteau, R. and Karp, F. (1979) *Arch. Biochem. Biophys.*, **198**, 512.

Dauner, H. and Mueller, G. (1975) *Hoppe-Seyler's Z. Physiol. Chem.*, **356**, 1353.

Deeg, R., Knemler, H.-P., Bergmann, K.-H. and Mueller, G. (1977) *Hoppe-Seyler's Z. Physiol. Chem.*, **358**, 339.

DeSilva, K. T., Smith, G. N. and Warren, K. E. H. (1971) *Chem. Commun.*, 905.

Dimroth, P., Ringlemann, E. and Lynen, F. (1976) *Eur. J. Biochem.*, **68**, 581.

Evans, R. and Hanson, J. R. (1976) *J. Chem. Soc., Perkin Trans, 1*, 326.

Fawcett, P. A., Usher, J. J., Huddleston, J. J. and Abraham, E. P. (1976) *Biochem. J.*, **157**, 651.

Gambliel, H. and Croteau, R. (1982) *J. Biol. Chem.*, **257**, 2335.

Gotfredsen, S. E. (1978) *The Biosynthesis of Lagopodine A, B, and C, Coccinol and Linalool*, Dissertation ETH (Zurich), No. 6243.

Gotfredsen, S., Obrecht, J. P. and Arigoni, D. (1977) *Chimia*, **31**,(2), 62.

Hendrickson, J. B. (1959) *Tetrahedron*, **7**, 82.

Hofschneider, P. H. and Goebel, W. (eds) (1982) *Current Topics in Microbiology and Immunology. Gene Cloning in Organisms Other than E. coli*, Springer-Verlag, Berlin.

Hopwood, D. A. and Merrick, M. J. (1977) *Bacteriol. Rev.*, **41**, 595.

Ikeda, H., Tanaka, H. and Omura, S. (1982) *J. Antibiot.*, **35**, 507.

Kohsaka, M. and Demain, A. L. (1976) *Biochem. Biophys. Res. Commun.*, **70**, 465.

Konomi, T., Herchen, S., Baldwin, J. E., Yoshida, M., Hunt, N. A. and Demain, A. L. (1979) *Biochem. J.*, **184**, 427.

Laskovics, F. M. and Poulter, C. D. (1981) *Biochemistry*, **20**, 1893.

Lee, S. L., Hirata, T. and Scott, A. I. (1979) *Tetrahedron Lett.*, 691.

Lewis, N. G., Neier, R., Matcham, G. W. J., McDonald, E. and Battersby, A. R. (1979) *J. Chem. Soc., Chem. Commun.*, 541.

Loew, P. and Arigoni, D. (1968) *Chem. Commun.*, 137.

Lynen, F., Eggerer, H., Henning, U. and Kessel, I. (1958) *Angew. Chem.*, **70**, 738.

Lynen, F., Agranoff, B. W., Eggerer, H., Henning, U. and Moeslein, E. M. (1959) *Angew. Chem.*, **71**, 657.

Mackie, H. and Overton, K. H. (1977) *Eur. J. Biochem.*, **77**, 101.

Mash, E. A., Gurria, G. M. and Poulter, C. D. (1981) *J. Am. Chem. Soc.*, **103**, 3927.

Meeschaert, B., Adriaens, P. and Eyssen, H. (1980) *J. Antibiot.*, **33**, 722.

Mizukami, H., Nordlov, H., Lee, S.-L. and Scott, A. I. (1979) *Biochemistry*, **18**, 3760.

Mombelli, L., Nussbaumer, C., Weber, H., Mueller, G. and Arigoni, D. (1981) *Proc. Natl. Acad. Sci. U.S.A.*, **78**, 11.

Mueller, G., Deeg, R., Gneuss, K. D., Gunzer, G. and Kriemler, H.-P. (1979a) in *Vitamin B_{12}* (eds B. Zagalak and W. Friedrich), Walter de Gruyter, Berlin, pp. 279–291.

Mueller, G., Gneuss, K. D., Kriemler, H.-P., Scott, A. I. and Irwin, A. J. (1979b) *J. Am. Chem. Soc.*, **101**, 3655.

Nozoe, S. and Machida, Y. (1972) *Tetrahedron*, **28**, 5105.

Ohfune, Y., Shirahama, H. and Matsumoto, T. (1976) *Tetrahedron Lett.*, 2869.

Omura, S. Ikeda, H. and Tanaka, H. (1981) *J. Antibiot.*, **34**, 478.

O'Sullivan, J., Bleaney, R. C., Huddleston, J. A. and Abraham, E. P. (1979) *Biochem. J.*, **184**, 421.

Overton, K. H. and Picken, D. J. (1976) *J. Chem. Soc., Chem. Commun.*, 105.

Parker, W., Roberts, J. S. and Ramage, R. (1967) *Q. Rev.*, **21**, 331.

Popjak, G. and Cornforth, J. W. (1960) *Adv. Enzymol. Relat. Subj. Biochem.*, **22**, 281.

Popjak, G. and Cornforth, J. W. (1966) *Biochem. J.*, **101**, 553.

Rasetti, V., Pfaltz, A., Kratky, C. and Eschenmoser, A. (1981) *Proc. Natl. Acad. Sci. U.S.A.*, **78**, 16.

Rueffer, M., Kan-Fan, C., Husson, H.-P., Steockigt, J. and Zenk, M. H. (1979) *J. Chem. Soc., Chem. Commun.*, 1016.

Ruzicka, L. (1959) *Proc. Chem. Soc. (London)*, 341.

Ruzicka, L. (1963) *Pure Appl. Chem.*, **6**, 482.

Ruzicka, L. and Stoll, M. (1922) *Helv. Chim. Acta*, **5**, 923.

Samuelsson, B. (1981) *Harvey Lect.*, **75**, 1.

Samuelsson, B., Goldyne, M., Granstrom, E., Hamberg, M., Hammarstrom, S. and Malmsten, C. (1978) *Annu. Rev. Biochem.*, **47**, 997.

Schleif, R. F. and Favreau, M. A. (1982) *Biochemistry*, **21**, 778.

Scott, A. I. (1970) *Acc. Chem. Res.*, **3**, 151.

Scott, A. I. (1975) *Tetrahedron*, **31**, 2639.

Scott, A. I. and Lee, S.-L. (1975) *J. Am. Chem. Soc.*, **97**, 6906.

Scott, A. I. Townsend, C. A., Okada, K., Kajiwara, M. and Cushley, R. J. (1972) *J. Am. Chem. Soc.*, **94**, 8269.

Scott, A. I., Yagen, B. and Lee, E. (1973) *J. Am. Chem. Soc.*, **95**, 5761.

Scott, A. I., Beadling, L. C., Georgopapadakou, N. M. and Subbarayan, C. R. (1974) *Biorg. Chem.*, **3**, 238.

Scott, A. I., Lee, S.-L., de Capite, P., Culver, M. G. and Hutchinson, C. R. (1977) *Heterocycles*, **7**, 979.

Scott, A. I., Irwin, A. J., Siegel, L. M. and Shoolery, J. N. (1978) *J. Am. Chem. Soc.*, **100**, 316.

Seto, H. and Yonehara, H. (1980) *J. Antibiot.*, **33**, 92.

Sinsheimer, R. L. (1977) *Annu. Rev. Biochem.*, **46**, 415.

Stephan, K. (1898) *J. Prakt. Chem.*, **58**, 109.

Stoeckigt, J. (1978) *J. Chem. Soc., Chem. Commun.*, 1097.

Stoeckigt, J. and Zenk, M. H. (1977) *J. Chem. Soc., Chem. Commun.*, 646.

Stoeckigt, J., Treimer, J. and Zenk, M. H. (1976) *FEBS Lett.*, **70**, 267.

Stoeckigt, J., Hoefle, G. and Pfitzner, A. (1980) *Tetrahedron Lett.*, **21**, 1925.

Tamm, Ch. and Breitensein, W. (1980) in *The Biosynthesis of Mycotoxins* (ed. P. S. Steyn), Academic Press, New York, pp. 69–104.

Uzar, H. C. and Battersby, A. R. (1982) *J. Chem. Soc. Chem. Commun.*, 1204.

Volpe, J. J. and Vagelos, P. R. (1973) *Annu. Rev. Biochem.*, **42**, 21.

Wang, Y.-G., Davies, J. E. and Hutchinson, C. R. (1982) *J. Antibiot.*, **35**, 335.

West, C. A. (1981) in *Biosynthesis of Isoprenoid Compounds* (eds J. W. Porter and S. L. Spurgeon), John Wiley, New York, pp. 375–411.

Wu, R. (ed.) (1979) *Methods Enzymol.*, **68**.

8 | The impact of enzymology in biochemistry and beyond

Keith E. Suckling

8.1 INTRODUCTION

Biochemistry, with its major component fields which include enzymology, occupies a central position in the spectrum of the physical and biological sciences, providing a link between them as well as a common ground on which the two major divisions of natural science may be practised together. The development of biochemistry has been and continues to be critically dependent on the application of physical and chemical methods to biological problems. The greater part of this book has considered how the ideas and concepts which have arisen out of just one area of biochemical study, that of enzymology, have been of value in stimulating work in other, mainly chemical, fields. We have seen that looking over the shoulder of the biologist has been an inspiration to many chemists, who hope to match and make use of the elusive and exquisite selectivity and catalytic power of enzymes.

To the biochemist, and more especially to the biologist, however, the study of enzymes is just a part of a much larger subject which comprises every kind of biological process. In one sense one could regard biochemistry as a sort of integrated enzymology since the great majority of biochemical processes depend upon enzymes for catalysis or upon other proteins with many enzyme-like properties. In biology it is important to study not only the catalytic mechanisms of enzymes but also how the rates of the reactions they catalyse are regulated. Enzymology is thus expanded into the context of the whole cell and organism. The wide scope of the subject is demonstrated by the fact that biochemists and their methods are to be found in almost any institution where biological work is conducted. Academic subject titles have no importance here; biochemistry (and therefore enzymology) is practised in buildings which may be designated Botany, Zoology, Genetics, Immunology, Clinical Chemistry, Surgery or Medicine. Such is the inventiveness of those who bestow titles on buildings and institutions that this list could be extended many times.

This wide diversification is evidence for the very great impact which biochemistry has had on biology as a whole. The implication of this fact must be that specialists in more classically biological and medical fields see that the techniques and concepts of biochemistry are particularly relevant to their field of study. We can find evidence for the importance of enzymological techniques by examining non-biochemical textbooks for comments such as the following:

'With few exceptions . . . progress in the study of human biochemical genetics has depended on increasingly sophisticated methods of analysing protein structure and quantity.' (Brock, 1978).

'The extensive use of enzyme activity determinations as aids to the diagnosis of disease in hospital laboratories all over the world is one of the most dramatic developments in modern medicine.' (Wilkinson, 1976).

'A particularly important branch of histochemistry is associated with the localisation of enzymes.' (Lewis, 1975).

However there are potential dangers for a subject whose concepts and methods have become so widely adopted as a basis for research by workers trained in other disciplines. The proper use of methods and concepts alike depends upon a clear understanding of the fundamental science on which they are based. Of course this problem is not unknown to biochemists themselves who may sometimes feel securely protected by their central position between physical, chemical and biological science. This is hinted at by the well-known saying that a biochemist is a scientist who talks about biology to chemists and about chemistry to biologists. (There are a number of versions of what biochemists talk about to each other.)

We can perhaps most clearly see the influence of enzymology on biochemistry by considering those areas which do not focus on enzymes and their molecular properties as their main objectives. The impact of enzymology on these areas, as on the whole of biochemistry, may take several forms which will be discussed in the following sections.

1. In a more general sense enzymology is part of a wider field of biochemical study, that of protein chemistry. The study of proteins with enzymic and other functions has developed in parallel, each influencing the other. To many chemists the study of enzymes has been predominant, but there is much important chemistry to be found in the study of non-enzyme proteins. Some of the wide range of proteins which can be studied by methods similar

Table 8.1 Non-enzymic protein systems

System	Function
Haemoglobin	Reversible binding of oxygen in blood (Perutz, 1979)
Antibody	Specific non-covalent binding to antigen (Amzel and Poljak, 1979)
Plasma lipoproteins	Binding of insoluble lipid molecules for transport in the blood (Smith *et al.*, 1978)
Calmodulin	Binds Ca^{2+}. The metal−protein complex initiates or regulates intracellular metabolism (Klee *et al.*, 1980)
Acetylcholine receptor	Binds acetylcholine released from a neuromuscular junction to initiate muscle contraction (Conti-Tronconi and Raftery, 1982)
Repressor protein	Interacts with a specific region of DNA in bacteria to regulate transcription (Ptashne *et al.*, 1982)
Cytoskeleton	Several proteins (e.g. tubulin) associate together and with the cell membrane to form a network within the cell (Timasheff and Grisham, 1980)
Endocytosis	The protein clathrin organizes regions of the plasma membrane for a cell at which vesicles form which allow the uptake of materials into the cell (Pearse and Bretscher, 1981)

to those used for enzymes are collected together in Table 8.1. Enzyme-catalysed reactions themselves are important in more complex physiological processes such as muscle contraction (Adelstein and Eisenberg, 1980), blood clotting (Jackson and Nemerson, 1980) and the removal of foreign cells from the blood by the complement system (Reid and Porter, 1981). As an illustration of the similarity between enzyme studies and the investigation of the function of proteins without catalytic activity we shall examine some aspects of the chemistry of the plasma lipoproteins at the end of Section 8.2.

2. Enzymes themselves may be used as tools as, for example, in clinical chemistry. Hybrid methods in which enzymes are coupled to other techniques such as electron microscopy and immunology have been developed (Section 8.3).

3. The most successful concepts derived from the study of enzymes may be applied to other systems. Such a transfer of concepts can be extremely important in stimulating new areas of research. The concepts of protein structure, conformation and the binding of small molecules to specific sites on a protein are widely applicable to non-enzyme systems. Again paradoxically many important concepts in enzymology were developed in the study of haemoglobin (Table 8.1), a protein which has been referred to as an 'honorary enzyme'. As we shall see later (Section 8.4), problems can arise when the concepts derived from the study of enzymes in solution are applied directly to more complex systems.

4. In the last section of this chapter we shall move from a mainly historical discussion to see how enzymology may be expected to develop and to influence biochemistry in the future. Earlier chapters in this book have considered the future implications of enzyme chemistry in more chemically oriented areas, but advances in the wider sphere of biology should not be ignored. There are many areas of biology where the time is now ripe for a detailed molecular study to be attempted and where the present power and adaptability of molecular enzymology may be tested. In some of these the concepts of what we might call 'small substrate molecule' enzymology may prove to be limiting. The increasingly well-characterized proteins whose substrates are macromolecules such as nucleic acids can now be studied in detail. More complex topological considerations are necessary here because of the complex folding of the nucleic acid molecules. Proteins which have so far been studied in purified form in dilute aqueous solution may experience quite a different environment within a cell and the significance of the protein–protein interactions which may result needs evaluating. Many proteins are constrained in partially hydrophobic environments in membranes, and although the basic concepts of membrane structure of the last decade have led to great progress, a good molecular understanding of none of these systems is available (Section 8.4). Fields which may appear superficially to be completely non-biochemical in character such as taxonomy and evolution have also gained new insights using ideas derived from enzyme structure and function.

Techniques, too, continue to develop. The application of magnetic resonance techniques to the study of protein structure and enzyme mechanism has been a conspicuous success of recent years (see, for example, Chapter 6). As we have already seen, a deeper understanding of the catalytic mechanism of enzymes has allowed more subtle and specific probe molecules to be designed. Even more subtle will be the modifications which can now be made in the structure of a protein by genetic engineering. Analytical techniques for the separation of macromolecules on a small scale and for the sequencing of proteins and nucleic acids are continually improving and this will allow the study of the molecular detail of important systems which are for various reasons inaccessible to established techniques.

In the remaining pages of this book we will examine in more detail some of the points raised in this opening discussion by illustrating some of the successes in biochemical areas which have developed in parallel to the study of enzymes and which, at a suitable stage in their development, have been aided by the physical and chemical techniques and concepts which are characteristic of enzymology. Little of what follows will be new to biochemists but chemists may find a broadening of the horizons at the end of this book to be a fitting open-ended close.

8.2 NON-ENZYME PROTEINS

The rapid and specific catalysis achieved by enzymes has been a major factor in capturing the attention and imagination of chemists, who by the nature of their subject are especially interested in chemical bonding and how these bonds can be made and broken. It is clear that the specificity of enzymes depends on their ability to recognize and to interact preferentially with their substrates as a prelude to and during the chemical reaction. Specificity in recognizing a small molecule or part of a large molecule is, however, a property of a large number of proteins and is not unique to enzymes as was realized by an enlightened few following Fischer (Chapter 1). Many biological processes other than catalysis of chemical reactions require similar precise recognition and this is usually provided by a protein with a specific binding site for the molecule concerned. Binding may be followed by other events through which a biological response is elicited. Table 8.1 presents a list of some biological systems in which precise molecular interactions involving proteins are essential but in which no chemical reactions are catalysed.

Studies of all these systems form part of the general field known as protein chemistry, and the study of non-enzyme proteins has developed in parallel to that of enzymes in this field. The different biological nature and functions of proteins without enzymic activity often leads to additional problems in their study. One important example is in methods for the assay (or measurement of the quantity) of a given protein. The presence of an enzyme is relatively easily detected by measuring the rate of the reaction which it catalyses. The only response that can be measured with many isolated non-enzyme proteins may be the binding of the appropriate ligand. Interactions such as these in which no chemical reaction takes place are much harder to measure with sensitivity. Methods for measurements of this kind include estimating the change in the fluorescence emission or in the circular dichroism spectrum of the protein that accompanies binding. Such effects can often be correlated with a conformational change in the protein which takes place when it interacts with its ligand (see below and Freifelder, 1982). Binding can also be measured physically by separating the protein with ligand bound to it from the unbound ligand by ultrafiltration, dialysis or specific absorption. Other proteins may be recognized only by a highly distinctive property such as an unusual absorption spectrum or a characteristic mobility on gel electrophoresis. A successful study of many of the systems outlined in Table 8.1 depends critically on the availability of a suitable assay.

Perhaps the similarities and contrasts in structural and functional studies of enzyme and non-enzyme proteins are best illustrated by an example. Fat is a major component of the diet and also a major energy store for the body. Since fat (chemically, triacylglycerol) is highly insoluble in water the body must provide a mechanism by which fat can be transported in the blood between those tissues which can store it and those which oxidize it for energy. This is achieved by the association of the fat with specific plasma proteins to form large polymolecular aggregates or particles with a defined structure. These particles are known as plasma lipoproteins since they contain both lipid and protein. They have been intensively studied in recent years because of the relationships which appear to hold between the amounts of the lipo-proteins of different types circulating in the blood and the incidence of diseases of the circulation such as atherosclerosis (Goldstein and Brown, 1977). Many of the proteins which form part of these particles have been purified and their structures examined in order to understand how they are able to interact specifically with triacylglycerol and other lipids such as chol-esterol, cholesteryl esters and phospholipids, effectively solubilizing them in large quantity. Some of these proteins, or apolipoproteins as they are called, have an additional function. As well as participating in the overall structure of the lipoprotein they are recognized by receptors on the plasma membrane of certain cells and this recognition event is thought to initiate the process by which the whole particle is taken up into the cell and degraded (Pearse and Bretscher, 1981).

The first requirement for a molecular study of a protein is usually purifi-cation. This is not too difficult for lipoproteins since the combination of lipid and protein gives them densities which are very different from proteins alone. The higher the amount of the lipid in the particle, the less its density, so the particles can be separated from other proteins by ultracentrifugation. Also, since lipoproteins differ in their relative contents of lipid and protein, they can be fractionated into different classes (Table 8.2).

The distribution of apoproteins is characteristic of the type of lipoprotein. For example, high-density lipoprotein or HDL contains a major protein called apoA. This protein can be purified from the HDL fraction in suf-ficient amounts to allow its structure to be studied in detail. In some respects

Table 8.2 Human plasma lipoproteins

	Chylomicrons	*VLDL*	*LDL*	*HDL*
Density range	0.95	0.95−1.006	1.006−1.063	1.063−1.210
Major lipids	Dietary TAG	Endogenous TAG	Cholesterol and esters	Phospholipids cholesteryl esters
% Protein	2	10	25	50
Major apoproteins	A-I, B, C	B, C, E	B	A-I, A-II

Key: VLDL = very low-density lipoprotein; LDL = low-density lipoprotein; HDL = high-density lipoprotein. TAG = triacylglycerol (triglyceride).

this detailed examination has gone further than the study of many purified enzymes.

One of the first questions to answer with pure apoA available is how it is able to interact with lipid and yet not precipitate from aqueous solution. ApoA appears to interact strongly with phospholipids, a polar lipid class related in structure to triacylglycerol and a major structural component of biological membranes (see Fig. 8.3). Binding of phospholipid can be measured by changes in the circular dichroism and fluorescence spectrum of the protein, which reflect conformational changes as it adapts to the ligand. Methods have been developed which allow the prediction of the folding of a polypeptide chain from a knowledge of its amino acid sequence (Chou and Fasman, 1978). The primary sequence (i.e. amino acid sequence) of apoA is known and predictions of the folding of the polypeptide chain based on this sequence were made (Andrews *et al.*, 1976). The predicted structure is shown schematically in Fig. 8.1. We would expect to find that the structure shows a hydrophobic region for binding lipid and a hydrophilic region which can favourably interact with water. Figure 8.1 suggests that this may be achieved by the formation of a helical structure such that the amino acid residues on one side of the helices are mainly hydrophobic and those on the opposite side are mainly hydrophilic.

Given such structural data the enzymologist would proceed to attempt to identify the amino acid residues which are essential for catalysis as the first stage in studying the catalytic mechanism. A similar question can be asked here. It would be useful to know which parts of the apoprotein are essential for the binding of lipid. In this system a very precise chemical approach is available which is not usually open to enzymologists because most enzymes are larger, more complex structures than apoA. This approach is to synthesize chemically fragments of the lipoprotein to see if they spontaneously form the predicted helical structure.

Syntheses of this type have been carried out by a modified Merrifield method for a number of apoproteins (Sparrow and Gotto, 1980) and the lipid-binding domains in apoA-I, apoA-II and apoC [from LDL (low-density lipoprotein)] have been identified. The fragments of these proteins necessary to activate enzymes in the plasma have also been defined. In some cases these studies found lipid-binding regions which were not predicted by the calculations based on the primary sequence. The movement of lipids within the body is a complex dynamic process in which the tissues interact with the lipoproteins and in which lipid and protein molecules exchange between lipoprotein particles themselves. These structural investigations are therefore part of a much larger study which aims to describe the dynamic processes of lipid metabolism in the body.

This example shows that a protein with a simple function, that of binding lipid molecules, can be examined at great depth in molecular terms in a way

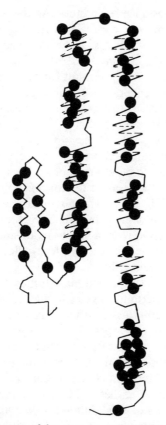

Fig. 8.1 Secondary structure of human apoA-I. The hydrophobic residues are shown by the black circles and can be seen to congregate close to each other in the nine helical regions of the protein. These are thought to provide the lipid-binding properties of the apoprotein. (Drawing based on Andrews *et al.*, 1976.)

very similar to studies of enzymes. This and much other work shows that the power of protein chemistry allows molecules with functions that are very different from enzymes to be studied with an equal expectation of success.

8.3 USES OF ENZYMES AS BIOLOGICAL TOOLS

We have seen earlier in this book how enzymes can be used in chemical synthesis and related activities. The major feature of enzymes which was attractive here was the high specificity that could be obtained. This property is also important in the analysis of small molecules in the complex mixtures that are often found in biological fluids. The selectivity of enzymes is such that measurements can often be made without the need for a preliminary purification of the substance of interest. As the following extracts show, applications of this type were appreciated early by clinical scientists.

'These applications' (enzyme assays) 'constituted only a small fraction of the total clinical work . . .' (in the 1930s) 'Today in the larger hospital laboratories enzyme assays may account for as much as 25 to 35 per cent of the total work load.' (Kachner, 1970). This is probably an underestimate for the 1980s. Whilst agreeing with this view, Schmidt (1979) writes that 'there are serious grounds for concern that a large proportion of this flood of measurements is useless.'

It is now routine for clinical laboratories to carry out many assays of small metabolites using enzyme-based methods which are highly automated. Common examples are the assay of the concentration of glucose or of cholesterol in the blood (Fig. 8.2). These two estimations of quite different compounds work on the same principle. The glucose or cholesterol is oxidized by an enzyme derived from a micro-organism, and hydrogen peroxide is formed. This in turn oxidizes a dye to a coloured form which can be measured spectrophotometrically.

Enzymes derived from the patient himself may also be important in diagnosis. The presence of certain enzymes in the blood is a good indication of specific tissue damage and this can often be characterized precisely by determining which of several tissue-specific forms is present. For example, raised levels of aminotransferases in the blood can be used as indices of a myocardial infarction. This observation, first made in 1954, was a major stimulus for the development of modern clinical chemistry (Wilkinson, 1976).

As the extracts quoted earlier pointed out, there are some important requirements if methods of this kind are to be reliable enough for clinical use. It is of paramount importance that the kinetic behaviour of the enzymes

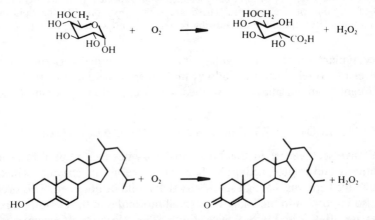

Fig. 8.2 The reactions catalysed by glucose oxidase (top) and cholesterol oxidase (bottom).

used in the assay is well understood under the reaction conditions to be used. If the rate of an enzyme-catalysed reaction is to be measured, it is essential that the substrates be present in concentrations which are saturating so that the measured rate is dependent only on the amount of enzyme present. If the concentration of a metabolite is to be measured using an enzyme, one must be sure that the time allowed for completion of the reaction is adequate over the range of concentrations to be measured. Clinically it is also important that enzyme-based analytical methods be standardized at a regional, national and international level both with regard to the method of assay and to the conditions under which the sample is to be obtained from the patient. Some of the greatest variability in assays of this type can be attributed not to the enzyme but to the methods of sampling and storage of the samples (Schmidt, 1979).

One of the most important advances in biological methods in the last decade has been the development of assays based on the specific binding that can be obtained with antibodies (Table 8.1). More recently the scope of these techniques has increased with the advent of monoclonal antibodies (Yelton and Scharff, 1981). Antibodies are released into the blood of an animal in response to a challenge by a large foreign molecule or a cell. They are proteins with a common general structure into which is incorporated a variable region which contains the binding site for a part of the specific foreign substance or antigen. The molecular nature of the antibody—antigen interaction has been studied in some detail (see Suckling and Suckling, 1980a, for a discussion). A family of methods has developed around antibodies for the measurement of very small quantities of biological molecules such as peptides and steroids. The most widely used technique is radioimmunoassay which depends solely on the binding of the antibody to the antigen. More recently methods have been developed to take advantage of the amplification of sensitivity that can be provided by enzymes (Engvall, 1980). The importance of the enzyme to these methods could not be more strongly put than by Engvall: 'enzymology is the key to success in EMIT' (enzyme multiplied immunoassay technique).

In EMIT a small molecule to be estimated is coupled covalently to an enzyme to form an enzyme conjugate. A suitable enzyme might be malate dehydrogenase because its activity can easily be estimated spectrophotometrically by following the absorbance changes due to changes in concentration of its coenzyme NADH. A fixed amount of the enzyme conjugate is incubated with a fixed amount of the antibody raised against the molecule to be estimated. Varying amounts of the unknown are added to a series of such incubations and these additions prevent the antibody from binding to the enzyme conjugate to a degree depending on the amount of the unknown which was added. In this way the amount of antibody bound to the enzyme conjugate varies with the concentration of the unknown. The activity of the enzyme is altered proportionately to the amount of bound antibody and thus also to the amount of the unknown that was added to the incubation. The binding effect is made more apparent by coupling it to the enzyme assay, a measurement which is quick, accurate and sensitive. The actual concentrations of the unknown will usually be read off against a calibration curve prepared under the same conditions. Compounds which can be analysed by EMIT include the thyroid hormone thyroxine and drugs such as barbiturates, gentamycin and theophylline.

There are a number of advantages of this technique over other enzyme-linked methods such as ELISA (enzyme-linked immunosorbent assay), in particular that no separation of the free antibody from the bound unknown is necessary. However, more substances can be assayed by the latter method (Engvall, 1980).

Many studies in cell biology require information on the fate of proteins in cells and in intact larger organisms. If a protein of interest could be labelled with a radioactive atom, its biochemical life could be followed throughout an animal and, by using microscopic methods, also within a cell. The most widely used procedure for preparing such a labelled protein is radioiodination. Several chemical techniques are available for this purpose but perhaps the mildest method available depends upon the use of enzymes. The enzyme lactoperoxidase will catalyse very selectively the iodination of tyrosine residues in a protein, a reaction akin to that catalysed by chloroperoxidase (Chapter 4). The radioiodine is incorporated into those tyrosine aromatic rings which are exposed on the surface of the protein (Morrison, 1980). Such radioiodinated proteins are useful not only in the study of their metabolic fate at the whole animal, tissue and subcellular levels but also in radioimmunoassays.

At a subcellular level another peroxidase, horseradish peroxidase, can be used as a tracer. Events such as pinocytosis, in which a cell takes up a small part of its external medium into an intracellular vesicle, can be studied in this way. The location of the peroxidase after uptake into a cell is found by electron microscopy. The enzyme is first allowed to react with a substrate to produce an electron-dense region in the sample at the sites where the enzyme is located. Coupling of the peroxidase to specific antibodies to cellular proteins of interest increases the range of systems which can be visualized at a cellular level using this enzyme (Lewis, 1975; Dempsey and Vernon, 1980a). Techniques of this type allow particular types of cells to be detected by their enzyme content. For example cholinergic neurones can be detected by the presence of cholinesterase. Enzymes can also be localized within a cell as, for example, acid phosphatase in lysosomes. A controlled use of these techniques can enable an estimate of the distribution of an enzyme activity within a cell to be made.

A further clinical use of enzymes is as therapeutic agents (Holcenberg,

1982). For example, aminohydrolases have been used in cancer chemotherapy to deplete the tumour cells of certain amino acids that are required for growth. This is a common tactic of cancer chemotherapy: here an enzyme rather than a small molecule is being used as a weapon.

The foregoing examples are just a small selection of areas of biology in which enzymes play an important role as tools. The study of enzymes has also given rise to many important concepts. For example, the simple concepts of enzyme kinetics are essential in the design of clinical assays which use enzymes. The molecular concepts which have arisen in enzymology also have affected the way in which biologists working in other areas think about their problems. We continue by looking at some examples of the way such concepts have influenced biological and biochemical practice.

8.4 IMPACT OF THE CONCEPTS OF ENZYMOLOGY

Enzymes are probably the best understood group of biological macromolecules at the molecular level. The concepts of protein structure, including the folding of polypeptide chains and the domain organization so produced (Rossman and Argos, 1981), active sites and specific binding and the importance of conformation and conformational change (Gutfreund, 1980) have all been developed in this field. The interaction of small molecules (substrates and inhibitors) with enzymes, which can often be detected by studying the kinetics of the reaction, forms the prototype for understanding many similar phenomena in non-enzyme systems. In some cases, for example the lipoproteins discussed earlier, the information available leads to a molecular description which is at least as good as that available for many enzymes. However, in most cases we have to content ourselves with arguing by analogy with the relatively few macromolecular systems which are well understood at the molecular level. Despite the great advances in enzymology over the last 30 years the number of such systems is not large (Warshel, 1981; Pincus and Scheraga, 1981). The 'classic' enzymes of biochemistry, chymotrypsin and lysozyme, and also haemoglobin may be included in this list.

Clearly it takes many years to reach a detailed understanding of a protein at the molecular level so biologists are inevitably forced into using the concepts derived from the 'classic' studies without having the evidence to support any proposals which may be made that would satisfy most accepted mechanistic criteria. For example, it is possible to discuss changes in enzyme activity caused by the binding of a ligand, or some similar event, by referring to a 'conformation change' of the protein (Suckling and Suckling, 1980b). Such changes are well known in many proteins, for example, hexokinase (Bennett and Steitz, 1978) and haemoglobin (Perutz, 1979).

Most commonly this concept is invoked to rationalize some observed behaviour, usually the results of kinetic experiments. The organic chemist

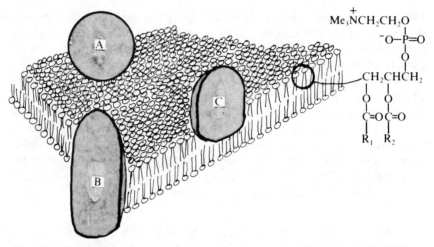

$$\overset{+}{Me_3}NCH_2CH_2O$$
$$|$$
$$^{-}O{-}P{=}O$$
$$|$$
$$O$$
$$|$$
$$CH_2CHCH_2$$
$$|\quad\quad|$$
$$O\quad\quad O$$
$$|\quad\quad|$$
$$C{=}O\ C{=}O$$
$$|\quad\quad|$$
$$R_1\quad R_2$$

Fig. 8.3 The fluid mosiac model of membrane structure. The phospholipid bilayer is associated with peripheral (or extrinsic) proteins bound only by ionic interactions (protein A) and with integral (or intrinsic) proteins which have regions which bind to the polar headgroup region of the membrane and the apolar core of the membrane. Protein B passes right through the phospholipid bilayer. Protein C penetrates just one half of it. The inset shows the structure of a typical phospholipid (phosphatidyl-choline) where R_1 and R_2 are fatty acyl groups.

who uses curly arrows to rationalize the behaviour of reacting molecules will be familiar with the limitations of such an approach (Suckling *et al.*, 1979). He may also be aware of the danger of the self-deception, easily succumbed to when powerful concepts such as these are used, that the processes under examination are being described reliably and in some depth. Analogies become weaker particularly rapidly when concepts are applied in fields which are remote from their origin.

In a related biochemical field, that of biomembranes, a similar situation arises in the use of a conceptually useful but imprecise mechanism to rationalize experimental results. Biomembranes consist of a bilayer of phospholipid molecules in which proteins are embedded (Fig. 8.3). In the simplest model the proteins are free to diffuse within the two-dimensional fluid formed by the bilayer. Clearly the rate at which this is possible will depend on the microviscosity of the fluid and this in turn is governed by such things as the fatty acids present in the lipids and the temperature. It is reasonable that a certain optimum fluidity may occur for the activity of a given integral membrane protein. This specific fluidity may be necessary, for example, to maintain the protein in its active conformation. The importance of the fluidity of the membrane is widely accepted (Lee, 1981), but as yet there are no well-understood systems at a molecular level which would allow one to link the physical state of the membrane directly with precise changes in the molecular structure of the protein in a way which would explain the observed activity. In such more complex biological systems some of the basic concepts of enzymology need to be tested for each system under study. We shall return to this problem shortly.

A further problem arises in the application of the concepts of enzymology to wider biological systems. Biochemists are not in complete agreement about what some of these simplest concepts are. Biochemical concepts, like those in many sciences, undergo changes as their experimental basis is tested. Often the meaning of a term is changed, usually in the direction of a more precise or specialized definition. The study of the kinetics of enzyme inhibition is a case in point. Most students become aware quite early in their studies of biochemistry that a variety of types of enzyme inhibition can be distinguished. These take names such as 'competitive', 'non-competitive' and the singularly uninformative 'uncompetitive'. The use of the term 'competitive' presumably carried with it from the first an implicit indication of the molecular mechanism of the inhibition. A natural interpretation would be that the inhibitor competed with the substrate for binding at the active site of the enzyme. As the understanding of kinetics and enzyme mechanisms has developed, biochemists have come to dispute the meaning of these terms.

To put these differences into perspective, a current treatment of the simple kinetics of enzyme inhibition might include the following. Two major types of inhibition can be distinguished, irreversible inhibition in which the enzyme becomes covalently bound to the inhibitor, and reversible inhibition. The family of competitive types of inhibition belong to the second group, that of reversible inhibition. A simple kinetic study of an enzyme involves the estimation of two basic parameters, $V_{max.}$, the maximum velocity at which the reaction can go, and K_m, the concentration of a given substrate at which half of this maximum velocity is reached. The classes of reversible inhibition can best be defined in terms of these two parameters. In competitive inhibition K_m is increased but $V_{max.}$ is not affected. With uncompetitive inhibition both K_m and $V_{max.}$ are decreased by the same factor. $V_{max.}$ is always decreased in non-competitive inhibition but K_m may change in either direction or not at all. The initial definition of the class of inhibition is obtained directly from the kinetics; a molecular interpretation may follow. This sequence of interpretation is central to many biochemical experiments and common to normal chemical practice (Chapter 2).

In one published view (Price, 1979) all one does in a kinetic study is to measure the rate of the enzyme-catalysed reaction under a variety of conditions. But the data obtained from such experiments do not allow one to make conclusions concerning the molecular structure of the enzyme and its relation to substrate and inhibitors. Kinetic studies do allow us to define the order in which substrate and inhibitors bind to the enzyme and to get some idea of whether two substrate molecules, or a substrate and an inhibitor,

can be bound to the enzyme at the same time. There may be structural implications in such results. This view does not attempt to approach the rather
precise molecular image conjured up by the term 'competitive'. It is quite
possible that the same kinetic data could have been obtained from a different
molecular interaction from the one traditionally implicit in 'competitive'.

This strict use of the terms as having only a kinetic meaning which reflects
the rate equations obtained for the reaction is not adhered to by some biochemists (Pace, 1980). Pace feels that a mechanistic definition is preferable,
despite the fact that this reverses the logic of the way in which a kinetic study
is carried out in practice. The more experimentally oriented view is dismissed
as unnecessary pedantry. However the International Union of Biochemistry
has adopted the former position (NC-IUB, 1982) and has also provided clear
definitions of kinetic constants for inhibition all couched entirely in
experimental terms.

In the kinetic experiments which are the subject of the preceding dispute,
the enzymes are usually studied in soluble form and the substrates and
inhibitors are all small water-soluble molecules. Diffusion of the molecules
can be assumed to be free in a three-dimensional solution and the concentrations of reactants can usually be known exactly. In these days of highly
purified enzymes the enzyme concentration may be known exactly too. Such
a defined system is clearly necessary for a meaningful kinetic study: enzymologists have often been warned against wasting 'clean thoughts on dirty
enzymes'. Returning to biological membranes, we can immediately see that
the situation is more complex. Here the proteins, if they diffuse at all, can
only diffuse with and within the membrane in which they are constrained.
The substrates for these enzymes may also be water-insoluble and conceivably may gain access to the enzyme active site by diffusion within the membrane. In this way both substrate and enzyme may be confined to a virtually
two-dimensional fluid with a viscosity very different from that of water.

The basic kinetic parameters, K_m and $V_{max.}$, are usually regarded as
properties of the enzyme. This view requires the assumption that all the
possible rate-limiting steps in the enzyme-catalysed reaction, beginning with
the binding of the substrate to the enzyme, are properties of the enzyme
itself. $V_{max.}$ and K_m can be measured for membrane-bound enzymes, but in
these more complex systems the situation is more uncertain. For example, we
do not know the concentration of the substrate as the enzyme sees it since the
substrate may partition between the aqueous solution and the membrane
phase to an unknown extent. We are no longer dealing with a homogeneous
solution. The kinetic parameters obtained for such a system will be valid
only for the conditions used in the experiment and will define the behaviour
of the whole experimental system. Either parameter may change if, for
example, a more efficient way of presenting the substrate to the enzyme were
devised. If diffusion of the substrate within the membrane is rate-limiting,

then the value of $V_{\text{max.}}$ will reflect this event rather than a catalytic property of the enzyme. A similar situation exists with immobilized enzymes used in industrial processes. Here K_m and $V_{\text{max.}}$ determined *in vitro* in solution may not relate at all to the kinetic properties of the enzyme once it is immobilized. Indeed, diffusion of the substrate on to the enzyme and of the product away from the enzyme may become rate-limiting in such processes. A different kinetic treatment is necessary to describe systems such as these (Goldstein, 1976).

8.5   RECENT AND FUTURE DEVELOPMENTS IN BIOCHEMISTRY

Biochemistry aims to define the events in living systems at a molecular level. This point seems trivial but it is worth emphasizing every so often to focus attention on the problems still to be solved (for example, Kornberg, 1980). Many of the early steps in biochemistry were taken in enzymology and the successes so obtained have depended on the application of a very wide range of methods, chemical, physical and biological in origin. Much of this book has been concerned with analysing the impact of enzymology on other fields of chemical science and in doing so only one aspect of a two-way process has been examined. Physical and chemical methods of purification, analysis and structure determination are the main tools of the enzymologist. If enzyme chemistry is now having a great impact on other areas of chemistry, we can regard this as the repayment of a debt to the fields from which the methodology was initially borrowed.

The exchange continues. Our grasp of molecular events improves with each new method. In recent years analytical techniques have become much more sensitive. Extremely small quantities of proteins or nucleic acids can now be purified and their sequences determined (Walsh *et al.*, 1981; Dempsey and Vernon, 1980b). The recent improvements in instrumentation in nuclear magnetic resonance allow the conformation and mobilities of macromolecules and their component domains to be determined in solution (Reid, 1981; Steitz and Shulman, 1982). These results can be compared with the picture obtained from the major structural method used in the solid, or more precisely, crystalline, state, X-ray diffraction. Nuclear magnetic resonance methods have also made possible the determination of the rates of enzyme-catalysed reactions in the intact cell (Gadian and Radda, 1981). Active sites of enzymes are also directly observable by this method (Cohn and Reed, 1982). In this way the knowledge already gained by the classical biochemical felony of breaking and entering the cell can be tested and the role of the intracellular milieu, which has so far been unobserved, may be evaluated.

Chemical studies on biological systems have become more subtle as knowledge has advanced and a conceptual basis for design of chemical attack has developed. The use of specific inhibitors and mechanistic probes has been discussed in Chapter 2. Chemical modification of proteins, an already widely used approach in the study of the relationship between structure and mechanistic function, can be made even more subtle by the technique of genetic engineering (Abelson and Butz, 1980; Ulmer, 1983). It will be possible to synthesize a DNA molecule coding for a desired protein containing specific alterations in the amino acids at any desired site by a combination of chemical and biological methods. This will allow the properties of the laboratory-produced 'mutant' proteins to be studied and the effects of very small changes in critical areas of the protein structure to be observed. Recently it has been demonstrated that a single change in an amino acid in a protein caused by the alteration of a single base in the DNA is enough to transform the cells in which this occurs into cancer cells (reviewed by Logan and Cairns, 1982).

Many of the areas in which there is yet a relatively poor understanding at a molecular level involve the important biological events in which two or more macromolecules interact with each other. Examples of this sort of interaction occur during DNA synthesis and its transcription to RNA and also during the assembly of proteins on the ribosomes (Brimacombe et al., 1978; Wittmann, 1982; Wool, 1979). Figure 8.4 illustrates some of the topological considerations which are necessary in systems of this kind (Champoux, 1978; De Pamphilis and Wassarman, 1980). Before the chemical interactions required for covalent bond formation can take place, many changes in the conformation of the nucleic acid must occur. These changes require the breaking and formation of many non-covalent bonds as the giant DNA molecule and its associated proteins gyrate into the necessary reactive conformations. Since a higher level of molecular organization is present in these systems than in the relatively small proteins studied classically, the molecular picture of events will require a new means of describing a kind of macromolecular topological catalysis before the more traditionally understood processes can be discussed (Gellert, 1981).

Such macro-structural considerations are also necessary in discussing many other biological events. Protein synthesis at the ribosome has already been mentioned. In membrane biology proteins are often found to aggregate together into functional units such as those involved in electron transport. The coupling of this process through the means of proton transfer from one side of the membrane to another to the synthesis of ATP, the central occupation of bioenergetic study, has been under intensive study for many years (Cross, 1981; Wikstrom et al., 1981). It is becoming clear that the complex intracellular arrays of structural proteins exist which help maintain the shape of the cell, and, in specialized cells such as muscle (which is better

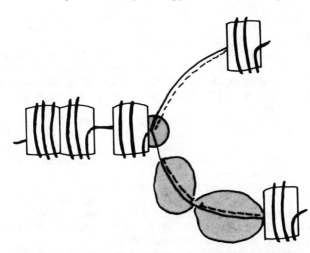

Fig. 8.4 The replication fork of a eukaryotic chromosome. The DNA is wound round the nucleosomes which consist of a core of proteins known as histones. Eight histone molecules are found in the nucleosome core and one between adjacent nucleosomes. The helix destabilizing protein promotes the dissociation of the DNA helix from the histone and its unwinding to form two single strands. Replication of the DNA takes place on both strands catalysed by DNA polymerase. The fragments produced on the lagging strand are joined by the DNA ligase. New nucleosomes are formed as histone molecules are made available. (Drawing based on DePamphilis and Wassarman, 1980).

understood), enable movement to take place. In areas such as these the conceptual framework provided by chemistry, which has sustained earlier studies, seems less secure. It may be that work in such biological areas will in turn influence chemical studies of non-biological molecules just as we have seen studies of enzymes influence chemistry so that the necessary scientific basis will be provided.

On the other boundary of biochemistry, in the wider biological area, the correlation of structure and function of proteins between different species has led to a broadened view in the field of evolution and speciation (Dickerson, 1980). Not only functional families but also structural families can be discerned in proteins (Rossman and Argos, 1981). The evolution and taxonomic significance of these comparisons has been stimulating to biology as a whole.

This is perhaps an appropriate example with which to close this chapter. We have seen how chemical, physical and biological methods and ideas have fused in the study of enzymes. The radiation produced by this fusion is penetrating into many fields of study, not just into chemical fields, as this book has amply demonstrated, but also to biology and science as a whole.

REFERENCES

Abelson, J. and Butz, E. (1980) *Science*, **209**, 1317.

Adelstein, R. S. and Eisenberg, E. (1980) *Annu. Rev. Biochem.*, **49**, 921.

Amzel, L. M. and Poljak, R. J. (1979) *Annu. Rev. Biochem.*, **48**, 961.

Andrews, A. L., Atkinson, D., Barratt, M. D., Finer, E. G., Hauser, H., Henry, R., Leslie, R. B., Owens, N. L., Phillips, M. C. and Robertson, R. N. (1976) *Eur. J. Biochem.*, **64**, 549.

Bennett, W. S. and Steitz, T. A. (1978) *Proc. Natl. Acad. Sci. U.S.A.*, **75**, 4848.

Brimacombe, R., Stoeffler, G. and Wittman, H. G. (1978) *Annu. Rev. Biochem.*, **47**, 217.

Brock, D. J. H. (1978) in *The Biochemical Genetics of Man* (eds D. J. H. Brock and O. Mayo), Academic Press, London, p. 4.

Champoux, J. J. (1978) *Annu. Rev. Biochem.*, **47**, 449.

Chou, P. Y. and Fasman, G. D. (1978) *Annu. Rev. Biochem.*, **47**, 251.

Cohn, M. and Reed, G. H. (1982) *Annu. Rev. Biochem.*, **51**, 365.

Conti-Tronconi, B. M. and Raftery, M. A. (1982) *Annu. Rev. Biochem.*, **51**, 491.

Cross, R. L. (1981) *Annu. Rev. Biochem.*, **50**, 681.

Dempsey, C. E. and Vernon, C. A. (1980a) *Annu. Rep. R. Soc. Chem.*, **77**, 340.

Dempsey, C. E. and Vernon, C. A. (1980b) *Annu. Rep. R. Soc. Chem.*, **77**, 333.

DePamphilis, M. L. and Wassarman, P. M. (1980) *Annu. Rev. Biochem.*, **49**, 627.

Dickerson, R. E. (1980) *Sci. Am.*, **242**, 98.

Engvall, E. (1980) *Methods Enzymol.*, **70**, 419.

Freifelder, D. (1982) *Physical Biochemistry*, W. H. Freeman, San Francisco.

Gadian, D. G. and Radda, G. K. (1981) *Annu. Rev. Biochem.*, **50**, 69.

Gellert, M. (1981) *Annu. Rev. Biochem.*, **50**, 879.

Goldstein, J. L. (1976) *Methods Enzymol.*, **44**, 397.

Goldstein, J. L. and Brown, M. S. (1977) *Annu. Rev. Biochem.*, **46**, 897.

Gutfreund, H. (1980) *Can. J. Biochem.*, **58**, 1.

Holcenberg, J. S. (1982) *Annu. Rev. Biochem.*, **51**, 795.

Jackson, C. M. and Nemerson, Y. (1980) *Annu. Rev. Biochem.*, **49**, 765.

Kachner, J. F. (1970) in *Fundamentals of Clinical Chemistry* (ed. N. W. Tietz), Saunders, Philadelphia, p. 362.

Klee, C. B., Crouch, T. H. and Richman, P. G. (1980) *Annu. Rev. Biochem.*, **49**, 489.

Kornberg, A. (1980) *Can. J. Biochem.*, **58**, 93.

Lee, A. G. (1981) *Nature (London)*, **294**, 695.

Lewis, P. R. (1975) *Adv. Opt. Electron Microsc.*, **6**, 171.

Logan, L. and Cairns, J. (1982) *Nature (London)*, **300**, 103.

Morrison, M. (1980) *Methods Enzymol.*, **70**, 214.

NC-IUB (1982) *Eur. J. Biochem.*, **128**, 281.

Pace, C. N. (1980) *Trends Biochem. Sci.*, **5**, 173.

Pearse, B. M. F. and Bretscher, M. S. (1981) *Annu. Rev. Biochem.*, **50**, 85.

Perutz, M. F. (1979) *Annu. Rev. Biochem.*, **48**, 327.

Pincus, M. R. and Scheraga, H. A. (1981) *Acc. Chem. Res.*, **14**, 299.

Price, N. C. (1979) *Trends Biochem. Sci.*, **4**, N272.

Ptashne, M. Johnson, A. D. and Pabo, C. O. (1982) *Sci. Am.*, **247**, 106.

Reid, B. R. (1981) *Annu. Rev. Biochem.*, **50**, 969.

Reid, K. B. M. and Porter, R. R. (1981) *Annu. Rev. Biochem.*, **50**, 443.

Rossman, M. G. and Argos, P. (1981) *Annu. Rev. Biochem.*, **50**, 497.

Schmidt, E. (1979) *Advances in Clinical Enzymology*, Karger, Basel, p. 2.

Smith, L. C., Pownall, H. J. and Gotto, A. M., Jr. (1978) *Annu. Rev. Biochem.*, **47**, 751.

Sparrow, J. T. and Gotto, A. M., Jr. (1980) *Ann. New York Acad. Sci.*, **348**, 187.
Suckling, C. J., Suckling, K. E. and Suckling, C. W. (1979) *Chemistry through Models*, Cambridge University Press, Cambridge.
Suckling, K. E. and Suckling, C. J. (1980a) *Biological Chemistry*, Cambridge University Press, Cambridge, Chapter 16.
Suckling, K. E. and Suckling, C. J. (1980b) *Biological Chemistry*, Cambridge University Press, Cambridge, Chapter 13.
Timasheff, S. N. and Grisham, L. M. (1980) *Annu. Rev. Biochem.*, **49**, 565.
Steitz, T. A. and Shulman, R. G. (1982) *Annu. Rev. Biophys. Bioeng.*, **11**, 419.
Ulmer, K. M. (1983) *Science*, **219**, 666.
Walsh, K. A., Ericsson, L. H., Parmelee, D. C. and Titani, K. (1981) *Annu. Rev. Biochem.*, **50**, 261.
Warshel, A. (1981) *Acc. Chem. Res.*, **14**, 284.
Wikstrom, M., Krab, K. and Saraste, M. (1981) *Annu. Rev. Biochem.*, **50**, 623.
Wilkinson, J. H. (1976) in *The Principles and Practice of Diagnostic Enzymology*, Arnold, London, p. vii.
Wittmann, H. G. (1982) *Annu. Rev. Biochem.*, **51**, 155.
Wool, I. G. (1979) *Annu. Rev. Biochem.*, **48**, 719.
Yelton, D. E. and Scharff, M. D. (1981) *Annu. Rev. Biochem.*, **50**, 657.

Index

First published 1984 by
Chapman and Hall Ltd
11 New Fetter Lane, London EC4P 4EE
Published in the USA by
Chapman and Hall
733 Third Avenue, New York NY 10017
© *1984 Chapman and Hall*

Typeset by Scarborough Typesetting Services
and printed in Great Britain by
J. W. Arrowsmith Ltd, Bristol

ISBN 0 412 25850 1

British Library Cataloguing in Publication Data

Enzyme chemistry.
  1. Enzymes
  I. Suckling, Colin J.
  547.7′58   QP601

  ISBN 0-412-25850-1

Library of Congress Cataloging in Publication Data

Main entry under title:

Enzyme chemistry.

  Bibliography: p.
  Includes index.
  1. Enzymes − Addresses, essays, lectures. 2. Enzymes − Industrial
applications − Addresses, essays, lectures.
I. Suckling, Colin J., 1947−
QP601.E5157   1984   574.19′25   83-25179
ISBN 0-412-25850-1

Enzyme Chemistry

Impact and applications

Edited by

Colin J. Suckling

*Royal Society Smith and Nephew Senior Research Fellow
and Reader in Organic Chemistry,
University of Strathclyde, UK*

LONDON   NEW YORK

CHAPMAN AND HALL